McGraw-Hill's

Math

GRADE 5

Mc
Graw
Hill
Education

New York   Chicago   San Francisco   Athens   London   Madrid
Mexico City   Milan   New Delhi   Singapore   Sydney   Toronto

4 5 6 7 8 9 10 11 12 13 14 15 DOW 20 19 18 17 16

ISBN   978-0-07-177558-8
MHID   0-07-177558-7

e-ISBN   978-0-07-177559-5
e-MHID   0-07-177559-5

Cataloging-in-Publication data for this title are on file at the Library of Congress.

Library of Congress Control Number: 2011915000

Printed and bound by RR Donnelley.

Editorial Services: Pencil Cup Press
Production Services: Jouve
Illustrator: Eileen Hine
Designer: Ella Hanna

McGraw-Hill Education products are available at special quantity discounts to use as premiums and sales promotions or for use in corporate training programs. To contact a representative, please visit the Contact Us pages at www.mhprofessional.com.

This book is printed on acid-free paper.

Table of Contents

Table of Contents

Welcome to McGraw-Hill's Math!

This book will help you succeed in your mathematics studies. Its short lessons explain key points and provide practice activities.

Look at the Table of Contents. It tells what topics are covered in each lesson. Seeing how a book is organized will help guide you as you work.

Then look at the 10-Week Summer Study Plan. You can use it to plan your time as you practice the skills in this book. Remember, the Summer Study Plan is only a guide. Work at your own pace.

Begin with the Pretest. This will help you determine whether you need to work on some skills more than others.

Each chapter ends with a Chapter Test. These post tests will show you what you have mastered as well as the skills that you may need to practice more.

A Posttest comes after the final chapter. You will solve problems from many of the lessons you have studied.

Remember to practice math. Practice will help you master the skills.

10-Week Summer Study Plan

Many students will use this book as a summer study program.
Use this 10-week study plan to help you plan your time.
Put a ✔ in the box when you have finished the day's work.

	Day	Lesson Pages	Test Pages
Week 1	Monday		PRETEST 8-13
	Tuesday	14, 15, 16	
	Wednesday	17, 18, 19	
	Thursday	20	21–22
	Friday	23, 24	
Week 2	Monday	25, 26	
	Tuesday	27, 28	
	Wednesday	29, 30	
	Thursday	31, 32	
	Friday	33	34–35
Week 3	Monday	36, 37	
	Tuesday	38, 39, 40	
	Wednesday	41, 42, 43	
	Thursday	44, 45, 46	
	Friday	47	48–49
Week 4	Monday	50, 51	
	Tuesday	52, 53, 54	
	Wednesday	55, 56, 57	
	Thursday	58	59–60
	Friday	61, 62, 63	
Week 5	Monday	64, 65	66–67
	Tuesday	68, 69	
	Wednesday	70, 71, 72	
	Thursday	73, 74	
	Friday	75, 76, 77	

	Day	Lesson Pages	Test Pages
Week 6	Monday	78	79–80
	Tuesday	81, 82	
	Wednesday	83, 84–85	
	Thursday	86–87, 88	
	Friday	89, 90–91	
Week 7	Monday	92, 93	94–95
	Tuesday	96, 97–98	
	Wednesday	99, 100–101	
	Thursday	102	103–104
	Friday	105, 106	
Week 8	Monday	107, 108	
	Tuesday	109, 110, 111	
	Wednesday	112, 113, 114	
	Thursday	115	116–117
	Friday	118, 119, 120	
Week 9	Monday	121, 122, 123	
	Tuesday	124, 125, 126	
	Wednesday	127, 128, 129	
	Thursday	130	131–132
	Friday	133, 134	
Week 10	Monday	135, 136	
	Tuesday	137, 138	
	Wednesday	139, 140	
	Thursday		141–142
	Friday		POSTTEST143–148

Name _____

Write each number in standard form.

1 53 and 125 thousandths

_____

2 96 thousandths

_____

3 2 billion, 109 million, 7

_____

Write each number in word form.

4 19,103,578,143

_____

5 200,200,200

_____

Write >, <, or = to compare each pair of numbers.

6 5.7 _____ 5.71

7 0.093 _____ 0.93

8 1.046 _____ 1.046

Round each number to the place of the underlined digit.

9 4.3$\underline{6}$9 _____

10 1.0$\underline{2}$5 _____

11 17.4$\underline{5}$7 _____

Add. Write the sum.

12
$$\begin{array}{r} 1.94 \\ + 7.59 \\ \hline \end{array}$$

13
$$\begin{array}{r} 557 \\ + 623 \\ \hline \end{array}$$

14
$$\begin{array}{r} 702 \\ + \ 99 \\ \hline \end{array}$$

15
$$\begin{array}{r} 5.38 \\ + 0.92 \\ \hline \end{array}$$

16 53,475 + 19,332 _____

17 12,257 + 8,153 _____

18 71.17 + 6.76 _____

Subtract. Write the difference.

19
$$\begin{array}{r} 713 \\ - \ 89 \\ \hline \end{array}$$

20
$$\begin{array}{r} 8.89 \\ - 2.98 \\ \hline \end{array}$$

21
$$\begin{array}{r} 176 \\ - \ 83 \\ \hline \end{array}$$

22
$$\begin{array}{r} 343.81 \\ - \ 66.10 \\ \hline \end{array}$$

23 15,432 – 13,718 = _____

24 150.62 – 42.37 = _____

25 83.01 – 78.81 = _____

26 A potato weighs 7.72 ounces. An apple weighs 5.03 ounces. How many ounces heavier is the potato than the apple?

_____

Simplify and complete. Tell what property is represented.

27 $20 \times 8 = ($ _____ $\times 8) + (10 \times$ _____ $) = 80 +$ _____ $=$ _____

_____ Property

28 $(30 \times 4) \times 5 =$ _____ $\times (4 \times$ _____ $) = 30 \times$ _____ $=$ _____

_____ Property

Multiply. Write the product.

29	**30**	**31**	**32**
46	28	511	123
$\times\ \ 3$	$\times\ 14$	$\times\ \ \ 4$	$\times\ \ 75$

33 A bakery bakes 5-dozen loaves of bread. Each loaf sells for 2 dollars. How much money will the bakery earn if it sells every loaf of bread?

_____

34 $2{,}811 \times 10^2 =$ _____   **35** $61 \times 10^3 =$ _____   **36** $77.6 \times 10^1 =$ _____

Divide. Write the quotient.

37 $81 \div 3 =$ _____   **38** $14 \div 4 =$ _____   **39** $78 \div 6 =$ _____

Estimate. Then multiply or divide.

40 Estimate: $3.1 \times 4 =$ _____   **41** Estimate: $7.13 \times 5 =$ _____   **42** Estimate: $11.09 \times 6 =$ _____

Multiply: $3.1 \times 4 =$ _____   Multiply: $7.13 \times 5 =$ _____   Multiply: $11.09 \times 6 =$ _____

43 $19.2 \div 10^4 =$ **44** $872.3 \div 10^2 =$ **45** $16.6 \div 10^3 =$

Estimate: _____   Estimate: _____   Estimate: _____

Quotient: _____   Quotient: _____   Quotient: _____

46 $14.85 \times 10^3 =$ **47** $1.256 \times 10^4 =$ **48** $6.798 \times 10^3 =$

Estimate: _____   Estimate: _____   Estimate: _____

Product: _____   Product: _____   Product: _____

Simplify and solve. Show your work.

49 $(10 - 8)^3 + (6 \times 4) \times (4 - 2) + 10^3 =$

Simplify inside the parentheses: _____

Simplify exponents: _____

Multiply: _____

Add and subtract from left to right: _____

Final answer: _____

50 $4\{[2(2 + 5) + 2] - 9\} =$

Simplify inside the parentheses: _____

Simplify inside the brackets: _____

Simplify inside the braces: _____

$4\{[2(2 + 5) + 2] - 9\} =$ _____

51 $(4 \times 7) \div 2^1 =$

_____

52 $4 + (10 \times 4) - 4^2 =$

_____

Add or subtract. Give the answers in simplest terms. Change any improper fractions to mixed numbers.

53 $\frac{1}{4} + \frac{2}{4} =$ _____

54 $\frac{3}{8} + \frac{4}{8} =$ _____

55 $\frac{5}{6} - \frac{1}{6} =$ _____

56 $2\frac{1}{3} + \frac{3}{4} =$ _____

57 $\frac{5}{12} + \frac{1}{8} =$ _____

58 $\frac{7}{9} + \frac{11}{18} =$ _____

59 $1\frac{2}{7}$
$- \frac{1}{5}$
_____

60 $6\frac{1}{3}$
$+ 3\frac{5}{6}$
_____

61 $7\frac{1}{5}$
$+ 8\frac{1}{2}$
_____

62 $9\frac{1}{2}$
$- 1\frac{2}{3}$
_____

Find the equivalent fraction.

63 equivalent to $\frac{4}{6}$, denominator 24 _____

65 equivalent to $\frac{1}{3}$, denominator 36 _____

64 equivalent to $\frac{1}{2}$, denominator 42 _____

66 equivalent to $\frac{3}{8}$, denominator 32 _____

Multiply.

67 $16 \times \dfrac{5}{8} =$ _____

68 $18 \times \dfrac{2}{3} =$ _____

69 $\dfrac{3}{10} \times \dfrac{4}{5} =$ _____

Solve.

70 Some ferrets are about 18 inches long. A hamster is about $\dfrac{1}{3}$ as long. About how many inches long is a hamster?

_____

71 $\dfrac{1}{2}$ of the photos on Delia's digital camera are of friends and family members. $\dfrac{1}{5}$ of those photos are of her best friend Aiesha. What fraction of the total number of photos on Delia's camera is of Aiesha?

_____

72 A mother blue whale is $87\dfrac{2}{3}$ feet long. Her baby is $22\dfrac{5}{8}$ feet long. About how much longer is the mother than the baby? Use front-end estimation to estimate the difference.

_____

Multiply to find the area of each rectangle.

73 9 ft long $\times \dfrac{3}{4}$ ft wide

area = _____

74 8 m long $\times \dfrac{1}{2}$ m wide

area = _____

Find each quotient. Draw models to help. Use multiplication to check your answers.

75 $\dfrac{1}{4} \div 9 =$ _____

76 $\dfrac{1}{2} \div 5 =$ _____

77 $12 \div \dfrac{1}{8} =$ _____

Name _____

Solve.

78 A giraffe's tongue is about 45 centimeters long. What is the length of a giraffe's tongue in millimeters?

_____

79 Mrs. Jones is measuring the mass of 2 eggplants. One eggplant has a mass of 0.68 kilograms and the other has a mass of 0.45 kilograms. What is the difference of the masses of both eggplants in grams?

_____

80 A server pours mineral water from a 1-liter bottle into a glass. After pouring, there are 0.3 liters left in the bottle. In milliliters, how much water did the server pour?

_____

81 Good hand washing involves using soap and clean, warm water for at least 20 seconds. If Sheng washes his hands 9 times each day, how many minutes does he spend washing his hands in 2 days?

_____

82 The Tyrannosaurus Rex had a skull up to 5 feet long. An adult human's skull is about 7 inches long. How many inches longer is a Tyrannosaurus Rex skull than a human skull?

_____

Classify each angle as straight, right, obtuse, or acute.

83

84

85

_____   _____   _____

Count the cubes in each object. Write the total.

86

87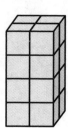

88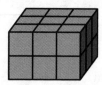

_____ cubes   _____ cubes   _____ cubes

Classify each triangle as equilateral, scalene, or isosceles. Then classify each triangle as acute, obtuse, or right.

89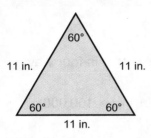

90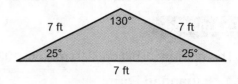

_____   _____

The heights of 6 ostriches are listed below. Complete the line plot to show this data set. Then use the line plot for Exercises 92 and 93.

91 Ostrich 1: $7\frac{1}{2}$ ft   Ostrich 2: $7\frac{3}{4}$ ft   Ostrich 3: 8 ft

Ostrich 4: 6 ft   Ostrich 5: $7\frac{1}{2}$ ft   Ostrich 6: $7\frac{1}{2}$ ft

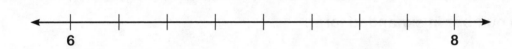

92 How much taller is the tallest ostrich than the shortest ostrich?

_____

93 What is the combined height of the ostriches with the most common height?

_____

Look at the grid. Write the ordered pair for each point.

94 *A* _____

95 *B* _____

96 *C* _____

97 *D* _____

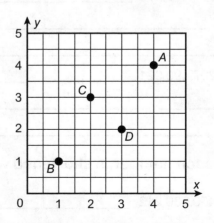

Lesson 1

Name _____

Place Value
You can write a number in different forms.

> ### Examples
>
> The number 1 billion, 210 million, 193 thousand, 422 can be written in four ways.
>
> 1. **Expanded Form:**
> $(1 \times 1{,}000{,}000{,}000) + (2 \times 100{,}000{,}000) + (1 \times 10{,}000{,}000) + (1 \times 100{,}000)$
> $+ (9 \times 10{,}000) + (3 \times 1{,}000) + (4 \times 100) + (2 \times 10) + (2 \times 1)$
>
> 2. **Word Form:**
>   one billion, two hundred ten million, one hundred ninety-three thousand, four hundred twenty-two
>
> 3. **Shortened Word Form:**
>   1 billion, 210 million, 193 thousand, **422**
>
> 4. **Standard Form:**
>   1,210,193,422

Write

Write each number in expanded form.

1 2,715,434,121 $\underline{(2 \times 1{,}000{,}000{,}000) + (7 \times 100{,}000{,}000) + (1 \times 10{,}000{,}000) + (5 \times 1{,}000{,}000)}$
$\underline{+ (4 \times 100{,}000) + (3 \times 10{,}000) + (4 \times 1{,}000) + (1 \times 100) + (2 \times 10) + (1 \times 1)}$

2 5,042,699,001 _____
_____
_____

Write each number in word form.

3 4,257,793,436 _____
_____
_____

4 6,207,490,021 _____
_____
_____

Write each number in shortened word form.

5 1,450,000,212 _____

6 5,842,776,481 _____

Patterns in Place Value

Recognizing patterns is a useful way to learn about numbers.

Examples

The value of 5 in the number 785 is different from the value of 5 in 458.

The **5** in 785 stands for 5 ones, or 5.

The **5** in 458 stands for 5 tens, or 50.

The 5 in 458 is one place to the **left** of the 5 in 785.
This makes the value of the 5 in 458 ten times greater than the value of the 5 in 785.

The 8 in 458 is one place to the **right** of the 8 in 785.

This makes the value of the 8 in 458 $\frac{1}{10}$ the value of the 8 in 785.

Solve

1 What is the value of 8 in 589? <u>80 or 8 tens</u>

What is the value of 8 in 438? _____

How does the value of 8 in 589 compare to the value of 8 in 438?

_____

2 What is the value of 7 in 367? _____

What is the value of 7 in 574? _____

How does the value of 7 in 367 compare to the value of 7 in 574?

_____

Describe the place value comparisons.

3 Compare the value of 4 in 4,605 to the value of 4 in 8,422.

_____

4 Compare 3 in 593 to 3 in 935.

_____

Solve.

5 Hal is thinking of two numbers. Each number includes the digits 3, 5, and 7. The 3 in the first number has $\frac{1}{10}$ the value of the 3 in the second number. The value of the 5 in the first number is 10 times greater than the 5 in the second number. The 7 is in the hundreds place in both numbers. What are the two numbers Hal is thinking of? _____

Name _____

Using Exponents

Another way to write a number is to use an exponent. An exponent is a number that tells how many times the base number is multiplied by itself.

Example

The exponent 3 in 10^3, shows that 10 is multiplied by itself three times: $10 \times 10 \times 10$.

Note that the exponent is written slightly smaller and higher than the base number.

To write 3,426 using exponents, you would write:
$(3 \times 10^3) + (4 \times 10^2) + (2 \times 10^1) + (6 \times 10^0)$

$$(3 \times 10^3) + (4 \times 10^2) + (2 \times 10^1) + (6 \times 10^0)$$
is equal to
$$(3 \times 10 \times 10 \times 10) + (4 \times 10 \times 10) + (2 \times 10) + (6 \times 0)$$

You can multiply with exponents:
$$23.56 \times 10^3 = 23.56 \times (10 \times 10 \times 10) = 23.56 \times 1,000 = 23,560$$

$$49,000 \div 10^3 = 49,000 \div (1,000) = \frac{49,000}{1,000} = 49$$

Sometimes the number 10^n is called a power of 10.

10^3 is the third power of 10.

Convert

Find each product or quotient. Show your work.

1. $75,600 \times 10^2$ $\underline{75,600 \times (10 \times 10) = 75,600 \times 100 = 7,560,000}$

2. $75,600 \div 10^2$ _____

3. 4×10^4 _____

4. $4 \div 10^4$ _____

5. $23 \div 10^3$ _____

6. 955×10^2 _____

7. 53×10^3 _____

8. What happens to the decimal point if you multiply a number by a power of 10?

_____

_____

Name _____

Place Value of Decimals

A decimal number has one or more digits to the right of the decimal point. Each digit to the left of the decimal point is a multiple of a power of 10. Each digit to the right of the decimal point is a dividend of a power of 10.

Examples

Here is the number 0.035 in a place value chart. The whole numbers are separated from the decimals by the decimal point.

Whole Numbers				Decimals		
hundreds	tens	ones		tenths	hundredths	thousandths
		0	.	0	3	5

There are several ways to write a decimal number.

1. Expanded form: $(0 \times 1) + (0 \times \frac{1}{10}) + (3 \times \frac{1}{100}) + (5 \times \frac{1}{1000})$

2. Word form: three **hundredths**, five **thousandths**

3. Shortened word form: 35 **thousandths**

4. Standard form: 0.035

Convert

Write in expanded form.

1. 35.726 ___$(3 \times 10) + (5 \times 1) + (7 \times \frac{1}{10}) + (2 \times \frac{1}{100}) + (6 \times \frac{1}{1,000})$___

2. 699.004 _____

Write in word form.

3. $[(4 \times 10) + (2 \times 1) + (5 \times \frac{1}{10}) + (7 \times \frac{1}{100}) + (6 \times \frac{1}{1,000})]$

_____

Write in standard form.

4. 5 and 401 thousandths _____   5. 24 thousandths _____

Write in shortened word form.

6. 1.451 _____   7. 1.001 _____

Name _____

Comparing Decimals Using Place Value

You can compare decimals by their place value.

Examples

A bass guitar string may be 0.831 mm thick or 0.839 mm thick. Which is thicker?

Compare these two decimal numbers to find which is greater. This will tell you which string is thicker.

Step 1: Write the numbers. Align the decimal points.

0.831
0.839

Step 2: Start from the left. Compare the digits until they are different.

0.831
0.839

9 is greater than 1. So 0.839 is greater than 0.831
0.839 > 0.831

Order these decimal numbers from smallest to greatest: 4.100, 4.353, 4.313

Step 1: Write the numbers. Align the decimal points. Add zeros if necessary.

4.100
4.353
4.313

Step 2: Start from the left. Compare the digits until they are different.

4.100
4.353   1 < 3
4.313

4.100 is the smallest of the three numbers.

Step 3: Continue to compare digits.

4.353   1 < 5
4.313   4.313 < 4.353

4.100 < 4.313 < 4.353

Compare

Write >, <, or = for each pair of numbers.

1 0.359 __<__ 0.421

2 0.4 _____ 0.004

3 3.212 _____ 32.101

Write the numbers in order from greatest to least.

4 0.305, 0.32, 0.345 _____

5 0.729, 7.29, 0.079 _____

6 0.456, 0.45, 0.4 _____

7 The chart shows the thickness of U.S. coins.

Coin	Cent	Nickel	Dime	Quarter	Half-Dollar
Thickness (cm)	0.155	0.195	0.135	0.175	0.215

List the coins in order from thickest to thinnest.

_____

Rounding Decimals
Rounding can be used to make decimals easier to work with.

Examples

Round 0.508 to the nearest hundredth.
You can use a number line.

0.508 is closer to 0.51 than to 0.50
0.508 rounded to the nearest hundredth is 0.51

You can also use a rule to round decimals.

Step 1: To round 0.436 to the nearest tenth, first identify the place you are rounding to.

0.436
↑
rounding place

Step 2: Look to the digit to its right.

0.436
↑
digit to the right

Step 3: If the digit to the right is 5 or greater, increase the rounding place digit by 1. If that digit is less than 5, do not change the rounding place digit. Drop all digits to the right of the rounding place digit.

0.436
3 < 5, so do not change the 4 and drop the digits to the right of 4.

0.436 rounded to the nearest tenth is 0.4

Round

Round each number to the place of the underlined digit. Plot each number on a number line if you need help.

1 6.5̲43 _____6.5_____

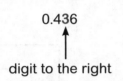

2 0.7̲42 _____

Round each number to the place of the underlined digit.

3 0.34̲5 _____

4 7.2̲99 _____

5 3̲.456 _____

6 3.55̲5 _____

7 0.98̲9 _____

8 9.09̲9 _____

Solve.

9 Sam buys a package of grapes marked 120.284 grams. How much is this to the nearest tenth of a gram?

_____

10 Ron makes acoustic guitars. He is looking for a string with a diameter of 0.0435 mm. What is the diameter of the string rounded to the nearest hundredth?

_____

Problem Solving: Logical Reasoning

You can use logical reasoning to solve some problems. First read the problem to understand the question. Then organize the data in a way that helps you think about the answer.

Example

Midge, Kat, Tomas, and Juan each have one backpack. Each is a different color. Midge's backpack is not green. Kat's backpack is not blue or white. Tomas's backpack is not blue. Juan's backpack is red.

First, list what you know.
- Each backpack is a different color.
- Kat's backpack is not blue or white.
- Juan's backpack is red.

- Midge's backpack is not green.
- Tomas's backpack is not blue.

Solve.
- Juan's backpack is red, so no other backpack can be red.
- Kat's backpack is not blue or white. It is also not red, so it must be green.
- This means that Midge and Tomas have backpacks that are either blue or white.
- But Tomas's backpack is not blue. So it must be white.
- The only color left is blue. Midge's backpack must be blue.

	Midge	Kat	Tomas	Juan
Red				yes
Green		yes		
Blue	yes			
White			yes	

Solution: Midge's backpack is blue, Kat's is green, Tomas's is white, and Juan's is red.

Solve

Use logical reasoning to find the answers.

1. Rafael, Jose, and Maria each went to one of the places shown in the table. Rafael's admission was not the most expensive. Maria's admission was less expensive than Rafael's. Where did each student go?

_____

_____

_____

City Museum Campus	
Location	Admission
Natural History Museum	$20.00
Aquarium	$25.95
Planetarium	$22.00

2. Alex, Brian, Candy, and Diana each competes in a different outdoor sport: skiing, skateboarding, rollerblading, or speed skating. Alex's sport does not need wheels or blades. Brian uses a skateboard. Candy's sport is not a cold-weather sport. In what sport does each person compete?

_____

_____

Name _____

1 What is the word form of 1,000,432,506?

_____

2 What does the 2 in 254 represent? _____

What does the 2 in 728 represent? _____

How does the value of 2 in 254 compare to the value of 2 in 728?

_____

3 How does the value of 9 in 5,935 compare to the value of 9 in 9,000? _____

4 What is 23×10^3 in expanded form? _____

What is 23×10^3 in standard form? _____

What is 23×10^3 in shortened word form? _____

Write the standard form of each number.

5 64×10^4 _____

6 $3,030 \div 10^3$ _____

7 0.5×10^4 _____

8 653.5×10^2 _____

9 $0.43 \div 10^1$ _____

10 0.77×10^2 _____

Write the shortened word form of each number.

11 6.92 _____

12 0.875 _____

13 72.7 _____

Compare the two numbers. Write <, >, or =.

14 23.00 _____ 2.300

15 2.454 _____ 2.456

16 11.117 _____ 11.17

17 0.3 _____ 0.300

18 4 _____ 0.004

19 9.9 _____ 10.01

Round to the nearest tenth.

20 23.271 _____

21 2.445 _____

22 1.117 _____

Round to the nearest hundredth.

23 23.271 _____

24 2.445 _____

25 1.117 _____

26 The average distance between the Earth and the Sun is 57 million kilometers. What is this number in standard form? _____

27 In 2011, the population of Mexico City was estimated to be 1<u>1</u>3,724,226. What is the value of the number to the left of the underlined digit?

_____

How does that value compare to the value of the underlined digit?

_____

28 In July 2009, the reported population of New York City was 8,391,<u>8</u>81. What is the value of the number to the right of the underlined digit?

_____

How does that value compare to the value of the underlined digit?

_____

Use the data in the chart for items 29 to 32.

29 What is the area of Tuvalu? Write the expanded form of this number.

_____

_____

30 What is the shortened word form for the area of Monaco?

_____

_____

Areas of Selected Small Countries	
Country	Area km^2
Tuvalu	23.309
Monaco	1.812
San Marino	62.159
Maldives	291.848
Marshall Islands	181.299

31 What is the area of San Marino in word form?

_____

_____

32 What is the area of the Marshall Islands rounded to the nearest hundredth?

_____

33 Four planes are lined up to take off. One will fly to Atlanta, one to Birmingham, one to Charlotte, and one to Denver. The plane to Birmingham is not the first or the last. The plane to Atlanta is third. The plane to Denver is not the first. In what order will the planes take off?

_____

Name _____

Addition Properties

Properties are rules that tell how numbers work. If you know how numbers work, you will know more than one way to solve some problems.

Examples

How can you use the properties of numbers to find the sums in these two addition exercises?

94 + 30 + 20 = ?

You can use the **Associative Property** and add the last two addends first, since they are easier to add.

94 + 30 + 20 = ?
94 + 50 = 144

74 + 62 = 136
62 + 74 = ?

How can you use properties to find the sum? If you add in a different order, using the **Commutative Property**, the sum will be the same.

62 + 74 = 136

Add

Use the Commutative Property to help you add.

1 54 + 73 = _____127_____   **3** 53 + 78 = _____

73 + 54 = _____127_____   78 + 53 = _____

2 29 + 96 = _____   **4** 88 + 65 = _____

96 + 29 = _____   65 + 88 = _____

Add. Use the Commutative or Associative Properties to help if you need to.

5 97 + 42 + 30 = _____   **6** 44 + 57 + 44 = _____   **7** 93 + 31 + 4 = _____

8 36 + 85 + 21 = _____   **9** 74 + 57 + 40 = _____   **10** 61 + 79 + 16 = _____

11 Ken wants to find the sum of 26 + 83 + 60. Which two numbers would be easiest to add first? Why?

_____

_____

_____

Lesson 2

Relating Addition and Subtraction

Addition and subtraction are inverse operations. Inverse operations undo each other. You can use known sums to solve subtraction problems. You can also use known differences to solve addition problems.

Examples

$48 + 29 = 77$

How can you find the difference of $77 - 29$?

Addition and subtraction are inverse operations. Since $48 + 29$ equals 77, we know that $77 - 29$ equals 48.

$108 - 65 = 43$

How can you find the sum of $43 + 65$?

Addition and subtraction undo each other. Since $108 - 65$ equals 43, we know that $43 + 65$ must equal 108.

Add or Subtract

Use inverse operations to solve.

1 $17 + 48 = 65$

$65 - 48 = \underline{\quad 17 \quad}$

2 $39 - 18 = 21$

$21 + 18 = \underline{\qquad\qquad}$

3 $64 + 83 = \underline{\qquad\qquad}$

$147 - 83 = \underline{\qquad\qquad}$

4 $56 - 27 = 29$

$29 + 27 = \underline{\qquad\qquad}$

5 $79 + 40 = 119$

$119 - 40 = \underline{\qquad\qquad}$

6 $82 - 58 = \underline{\qquad\qquad}$

$24 + 58 = \underline{\qquad\qquad}$

7 $71 + 57 = 128$

$128 - 57 = \underline{\qquad\qquad}$

8 $79 - 41 = 38$

$38 + 41 = \underline{\qquad\qquad}$

9 $78 + 56 = \underline{\qquad\qquad}$

$134 - 56 = \underline{\qquad\qquad}$

10 Teddy wants to know the value of x in $49 + x = 56$. He knows the difference of 56 and 49 is 7. How can he find the value of x? Explain.

_____

_____

_____

11 Teresa wants to find the difference of 136 and 117. She knows the sum of 117 and 19 is 136. How can she find the difference? Explain.

_____

_____

_____

Adding Whole Numbers

The most common method to use when adding numbers is to add them by place values.

Example

Find 548 + 201.

Step 1: Line up the digits by place value.

$$\begin{array}{r} 548 \\ + \ 201 \\ \hline \end{array}$$

Step 2: Add each column.

$$\begin{array}{r} 548 \\ + \ 201 \\ \hline 749 \end{array}$$

548 + 201 = 749

Add

Solve.

1. $\begin{array}{r} 230 \\ + \ 59 \\ \hline 289 \end{array}$

2. $\begin{array}{r} 411 \\ + \ 73 \\ \hline \end{array}$

3. $\begin{array}{r} 763 \\ + \ 79 \\ \hline \end{array}$

4. $\begin{array}{r} 402 \\ + \ 47 \\ \hline \end{array}$

5. $\begin{array}{r} 199 \\ + \ 146 \\ \hline \end{array}$

6. $\begin{array}{r} 209 \\ + \ 781 \\ \hline \end{array}$

7. $\begin{array}{r} 435 \\ + \ 69 \\ \hline \end{array}$

8. $\begin{array}{r} 259 \\ + \ 487 \\ \hline \end{array}$

9. $\begin{array}{r} 18 \\ + \ 183 \\ \hline \end{array}$

10. $\begin{array}{r} 721 \\ + \ 109 \\ \hline \end{array}$

11. $\begin{array}{r} 343 \\ + \ 75 \\ \hline \end{array}$

12. $\begin{array}{r} 521 \\ + \ 112 \\ \hline \end{array}$

13. 2,109 + 181 _____

15. 3,674 + 669 _____

14. 35,129 + 24,761 _____

16. 73,795 + 5,227 _____

17. In 2010, the three most populated cities in Montana were Billings, 104,170; Missoula, 66,788; and Great Falls, 58,505. What was the total population of these three cities?

_____

Name _____

Subtracting Whole Numbers

The most common method to use when subtracting numbers is to subtract them by place value.

Example

Find 782 − 251

Step 1: Line up the digits according to place value.

$$\begin{array}{r} 782 \\ -\ 251 \\ \hline \end{array}$$

Step 2: Subtract. Make sure to subtract the correct place values.

$$\begin{array}{r} 782 \\ -\ 251 \\ \hline 531 \end{array}$$

782 − 251 = 531

Subtract

Solve.

1
$$\begin{array}{r} 445 \\ -\ 213 \\ \hline 232 \end{array}$$

2
$$\begin{array}{r} 331 \\ -\ 221 \\ \hline \end{array}$$

3
$$\begin{array}{r} 176 \\ -\ 45 \\ \hline \end{array}$$

4
$$\begin{array}{r} 319 \\ -\ 11 \\ \hline \end{array}$$

5
$$\begin{array}{r} 579 \\ -\ 447 \\ \hline \end{array}$$

6
$$\begin{array}{r} 209 \\ -\ 159 \\ \hline \end{array}$$

7
$$\begin{array}{r} 430 \\ -\ 145 \\ \hline \end{array}$$

8
$$\begin{array}{r} 742 \\ -\ 472 \\ \hline \end{array}$$

9
$$\begin{array}{r} 873 \\ -\ 105 \\ \hline \end{array}$$

10
$$\begin{array}{r} 663 \\ -\ 359 \\ \hline \end{array}$$

11
$$\begin{array}{r} 717 \\ -\ 94 \\ \hline \end{array}$$

12
$$\begin{array}{r} 135 \\ -\ 109 \\ \hline \end{array}$$

13 4,199 − 2,180 _____

15 5,003 − 774 _____

14 46,224 − 24,121 _____

16 53,072 − 4,227 _____

17 Pamela the elephant was 113 kg at birth. As an adult, her mass is 4,990 kg. How many kilograms did she gain?

_____

Name _____

Models for Adding Decimals
You can use models to help you add decimal numbers.

Examples

A recipe for bread calls for 1.40 grams of wheat flour, 0.75 grams of rye flour, and 1.60 grams of oat flour. What is the total amount of flour in the recipe?

You can use a number line:

0.40 in 1.40 for wheat flour and 0.60 in 1.60 for oat flour combine to equal 1 whole.

So 1.**40** + 1.**60** = 1 + 1 + 1 = 3.

Add 0.75 of rye flour to get a total of 3.75.

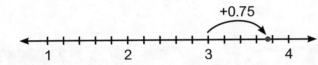

You can also use hundredths squares:

Break 1.40 into 1.00 + 0.40 and the 1.60 into 1.00 + 0.60. Combine the two 1s to get 2. Then combine the 0.40 and 0.60 to get another 1.00.

Add: 2 + 1 + 0.75 = 3.75.

1 + 1 = 2   0.40 + 0.60 = 1   0.70 + 0.05

Add

Draw number lines or grids to help you solve the following addition problems.

1 1.30 + 0.70 + 0.45 = ___2.45___

2 1.45 + 0.50 + 0.65 = _____

3 1.04 + 0.6 + 0.45 = _____

4 6.20 + 4.71 + 0.28 = _____

5 2.25 + 0.05 + 0.75 + 0.30 = _____

6 Ellen bought mushrooms for $1.49, beans for $1.19, and a dozen eggs for $2.32. Describe one way to find how much Ellen spent.

_____

_____

Name _____

Adding Decimals
You can use place value to add decimal numbers too.

Example

James bought three kinds of food for his pet fish. The packages were marked 9.51 ounces, 3.06 ounces, and 0.11 ounces. How many ounces of fish food did James buy in all?

Step 1: Line up the numbers at the decimal points.

```
   9.51
   3.06
 + 0.11
```

Step 2: Add each digit the same way you would add whole numbers.

```
   9.51
   3.06
 + 0.11
  12.68
```

James bought 12.68 ounces of fish food.

Add

1.
```
   2.13
 + 4.59
   6.72
```

2.
```
   4.01
 + 2.23
```

3.
```
   8.24
 + 2.76
```

4.
```
   9.03
 + 0.47
```

5.
```
   7.09
 + 3.23
```

6.
```
   1.11
 + 9.91
```

7.
```
   3.37
 + 1.56
```

8.
```
   2.83
 + 5.52
```

9. 19.90 + 4.32 _____

11. 4.22 + 0.78 _____

10. 23.88 + 1.75 _____

12. 8.77 + 9.12 _____

13.
```
   3.29
   2.81
 + 2.61
```

14.
```
  37.79
 173.45
+ 52.00
```

15.
```
   0.88
  10.45
 + 4.12
```

16. The Campos family spent the day fishing. They caught three brook trout. One fish weighed 12.45 pounds, one weighed 7.55 pounds, and one weighed 6.05 pounds. How many pounds of brook trout did they catch in all?

_____

Strategies for Adding Decimals

You learned several strategies for adding decimal numbers. Choose the strategy that works best for you.

Examples

You can use subtraction to help you add.

$17.51 - 6.87 = 10.64$
$10.64 + 6.87 = 17.51$

You can use properties of addition to help you add.

$11.81 + 20 + 9.73$
$11.81 + (20 + 9.73)$
$11.81 + 29.73 = 41.54$

You can use place value to line up the digits and decimal points before adding.

$$\begin{array}{r} 4.45 \\ 2.24 \\ +\ 1.20 \\ \hline 7.89 \end{array}$$

You can use number lines.

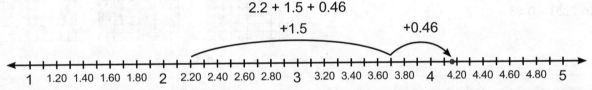

$2.2 + 1.5 + 0.46$

Add

Tell which strategy you used to solve each problem. Explain why you used it.

1. $1.14 + 3.06$ _____4.2_____

   Which strategy did you choose? _____

   Why? _____

2. $2.21 + 1.43$ _____

   Which strategy did you choose? _____

   Why? _____

3. $6.59 + 3.86$ _____

   Which strategy did you choose? _____

   Why? _____

4. $10.01 + 0.76$ _____

   Which strategy did you choose? _____

   Why? _____

5. $7.04 + 2.46$ _____

   Which strategy did you choose? _____

   Why? _____

Name _____

Models for Subtracting Decimals
You can use models to help you subtract decimal numbers.

Examples

A carton of milk contains 2.20 liters. Jake pours two glasses of milk, totaling 0.48 liters. How much milk is left in the carton?

You can use a number line.

0.48 can be written as 0.20 + 0.28.
Subtract 0.20 from 2.20, which leaves 2.00.
Then subtract 0.28 from 2.00 to get 1.72.

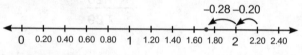

So 2.20 − 0.48 = 1.72.

You can also use hundredths squares.

Color 2.20 squares on 3 hundredths grids. Break **0.48** into **0.20 + 0.28** and cross out the 0.20. Then cross out the 0.28 on one of the fully colored hundredths grids.

2.20 − 0.48 = 1.72.

Subtract

Draw number lines or grids to help you solve these subtraction problems.

1 1.30 − 0.25 _____1.5_____

2 1.45 − 0.65 _____

3 2.59 − 1.09 _____

4 15.80 − 7.31 _____

5 2.55 − 0.30 _____

6 Bessie, a cow at Sunrise Farms, produced 20.16 kilograms of milk in one day. That morning, she produced 8.89 kilograms of milk. How much milk did Bessie produce in the evening?

_____

Subtracting Decimals

You can use place value to help you subtract decimal numbers.

Example

Anna has a small weather station outside her window. On Monday, the rainfall measured 2.21 inches. On Wednesday, the rainfall measured 4.62 inches. What is the difference in the two readings?

Step 1: Line up the numbers at the decimal points.

$$\begin{array}{r} 4.62 \\ -\ 2.21 \end{array}$$

Step 2: Subtract each digit the same way you would subtract whole numbers.

$$\begin{array}{r} 4.62 \\ -\ 2.21 \\ \hline 2.41 \end{array}$$

The difference in the two readings is 2.41 inches.

Subtract

Use place values when you solve the following subtraction problems.

1
$$\begin{array}{r} 4.13 \\ -\ 3.02 \\ \hline 1.11 \end{array}$$

2
$$\begin{array}{r} 5.55 \\ -\ 2.25 \end{array}$$

3
$$\begin{array}{r} 9.53 \\ -\ 3.44 \end{array}$$

4
$$\begin{array}{r} 19.34 \\ -\ 0.57 \end{array}$$

5
$$\begin{array}{r} 10.12 \\ -\ 8.33 \end{array}$$

6
$$\begin{array}{r} 15.99 \\ -\ 7.21 \end{array}$$

7
$$\begin{array}{r} 24.55 \\ -\ 22.78 \end{array}$$

8
$$\begin{array}{r} 17.19 \\ -\ 9.93 \end{array}$$

9 13.65 − 9.24 _____

11 4.22 − 0.78 _____

10 16.09 − 8.17 _____

12 22.14 − 10.07 _____

13
$$\begin{array}{r} 4.99 \\ -\ 3.33 \end{array}$$

14
$$\begin{array}{r} 8.25 \\ -\ 3.75 \end{array}$$

15
$$\begin{array}{r} 112.54 \\ -\ 21.45 \end{array}$$

16
$$\begin{array}{r} 18.55 \\ -\ 7.35 \end{array}$$

17 The chefs at Kim's Cafe made a total of 149.5 pounds of Korean ribs over a weekend. A wedding party ordered 35.66 pounds of ribs. How many pounds of ribs were left to sell?

_____

Strategies for Subtracting Decimals

You learned several strategies for subtracting decimal numbers. Choose the strategy that works best for you.

Examples

You can use addition to help you subtract.

$15.38 + 9.82 = 25.20$
$25.20 - 9.82 = 15.38$

You can use place value to line up the digits and decimal points before subtracting.

$$
\begin{array}{r}
8.56 \\
- \ 4.42 \\
\hline
4.14
\end{array}
$$

You can draw hundredth squares.

$9.52 - 7.31$

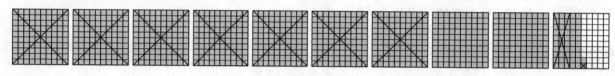

$9.52 - 7.31 = 2.21$

Subtract

Tell which strategy you used to solve each problem. Explain why you used it.

1. $3.59 - 2.06$ ___1.53___

   Which strategy did you choose? _____

   Why? _____

2. $2.21 - 1.23$ _____

   Which strategy did you choose? _____

   Why? _____

3. $6.59 - 3.49$ _____

   Which strategy did you choose? _____

   Why? _____

4. $10.01 - 9.99$ _____

   Which strategy did you choose? _____

   Why? _____

5. $72.04 - 12.40$ _____

   Which strategy did you choose? _____

   Why? _____

Problem Solving: Multistep Problems

Sometimes you need to complete more than one step to solve a problem.

Example

At Sal's Skateboard Shop, Ben bought the items shown on the receipt. He gave the cashier three $10 dollar bills. How much change did he get back?

Sal's Skateboard Shop	
wheels	$ 17.76
bushings	3.08
grip tape	4.73

Step 1: Decide what you need to know to solve the problem.
- the total cost of what Ben bought
- the total amount Ben gave the cashier

Step 2: Decide what operation or operations you will use to solve the problem.
- add to find the cost of the items bought
- add to find how much Ben gave the cashier
- subtract the cost from the amount Ben gave the cashier

Step 3: Find the solution.
$17.76 + $3.08 + $4.73 = $25.57
$10 + $10 + $10 = $30
$30.00 − $25.57 = $4.43
Ben received $4.43 back from the cashier.

Solve

1 A family of 2 adults and 2 children has $30.00 to spend on roller coaster rides. Do they also have enough money for each person to buy a Mexican grilled corn? Explain.

Roller Coaster Rides	
Adults	$ 7.75
Children	$ 4.50

Rios Food Stand
$ 0.85 each
Mexican Grilled Corn

Yes; $7.75 + $7.75 + $4.50 + $4.50 = $24.50 on roller coaster rides;

$0.85 + $0.85 + $0.85 + $0.85 = $3.40 for Mexican grilled corn. $24.50 + $3.40 = $27.90

2 Teri put four large bottles of water into a grocery bag. Each bottle weighs 2.19 pounds. In another bag, Teri put three cans of tomatoes. Each can weighs 1.75 pounds. Which bag weighs less?

_____

How much less? _____

3 JT has a small computer on his bike that tracks the number of miles he travels. Last Saturday, he traveled 4.23 miles to band practice, 3.72 miles to meet his friends, and 5.42 miles home. How many miles did he travel in all? _____

Name _____

Add.

1 322 + 353 = _____

2 75 + 176 + 25 = _____

3 42 + 34 = _____

  34 + 42 = _____

4 18.54 + 38.35 = _____

5 529 – 87 = 442

  442 + 87 = _____

6 456 + 247 = _____

7 396 + 700 + 70 = _____

8 564 + 374 = 938

  374 + 564 = _____

9 4.93 + 27.5 + 61.04 = _____

10 890 – 255 = 635

  635 + 255 = _____

Add. Draw number lines or grids to find the correct answer.

11 7.34 + 0.57 + 4.56 = _____

12 57.79 + 10.21 + 82.5 = _____

Add. Tell which strategy you used. Explain why you used it.

13 74.24 + 14.06 = _____

  Which strategy did you choose? _____

  Why? _____

Subtract.

14 584 – 273 _____

15 24.74 – 10.13 _____

16 375.07 – 88.12 _____

17 208 – 53 _____

18 67.21 – 25.9 _____

19 864.23 – 298.27 _____

Subtract. Draw number lines or grids to find the correct answer.

20 16.3 – 6.2 = _____

21 86.57 – 21.82 = _____

Subtract. Tell which strategy you used. Explain why you used it.

22 65.90 – 30.50 = _____

Which strategy did you choose? _____

Why? _____

Solve.

23 The Ruiz family, 2 adults and 2 children, visit the science museum. The Brinker family, 1 adult and 3 children, visit the zoo. Each family has $60.00 to spend on admission. Which family will spend less on admission?

_____

Price of Admission		
	Adult	**Child**
Science Museum	$15.00	$10.00
Zoo	$13.50	$9.50

24 A bird-watcher measured the heart rate of a blue-throated hummingbird at 1,260 beats per minute. She read that an inactive hummingbird may have a heart rate as low as 50 beats per minute. What is the difference between these heart rates?

_____

Name _____

Identity and Commutative Properties of Multiplication

You already know about properties that apply to addition. Some of the same properties also apply to multiplication.

Examples

A school lunchroom has 11 tables. If six students can sit at each table, how many students can eat lunch at the same time?

Find the answer by drawing 6 rows of 11 dots each.

You can also find the answer by drawing 11 rows of 6 dots each.

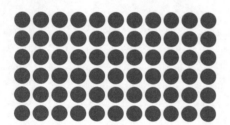

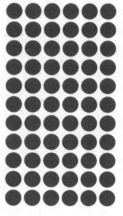

Multiplying 6 × 11 = 66.

Multiplying 11 × 6 = 66.

These solutions are examples of the **Commutative Property of Multiplication**.
If *a* and *b* are numbers, $a \times b = b \times a$.

The product of any number multiplied by 1, is that number. This is an example of the **Identity Property of Multiplication**.
If *a* is a number, $a \times 1 = a$.

Solve

1. Morgan draws two sketches to help solve a multiplication problem.

   What multiplication problem do the sketches represent?

   _____

   Which multiplication property do the sketches represent?

   _____

Find the product. Name the property illustrated.

2. 4 × 2 = 2 × 4 = _____

   _____ Property

3. 6 × 1 = _____

   _____ Property

4. 6 × 3 = _____

   _____ Property

5. 8 × 1 = _____

   _____ Property

Associative and Distributive Properties of Multiplication
There are two more properties that can help you solve multiplication problems.

Examples

Over the summer, Jake worked at his dad's store 3 hours a day, 5 days a week. He worked for 8 weeks. By the end of the summer, how many hours had Jake worked?

Here's one solution:
Multiply 3 hours/day × 5 days/week.
3 × 5 = 15 hours/week.
Then multiply 15 hours/week × 8 weeks.
15 × 8 = 120 hours
You can write this as (3 × 5) × 8 = 120.

Here's another solution:
Multiply 5 days/week × 8 weeks.
5 × 8 = 40 days
Then multiply 40 days × 3 hours/day.
40 × 3 = 120 hours
You can write this as 3 × (5 × 8) = 120.

Both of these solutions are examples of the **Associative Property of Multiplication**.
If a, b, and c are numbers, $(a \times b) \times c = a \times (b \times c)$.

There is a third way to solve the problem:
Multiplying 3 hours/day × 5 days/week = 15 hours/week.
Then break 15 into 1 ten and 5 ones and then multiply: $8 \times (10 + 5) = (8 \times 10) + (8 \times 5) = 120$.

This is an example of the **Distributive Property of Multiplication**.
If a, b, and c are numbers, $a \times (b + c) = (a \times b) + (a \times c)$.

Solve

Fill in the missing information. Name the property.

1. $(17 \times 5) \times 20 = \underline{\quad 17 \quad} \times (5 \times \underline{\quad 20 \quad}) = 17 \times \underline{\quad 100 \quad} = \underline{\quad 1700 \quad}$

   _____ Property

2. $11 \times 37 = (\underline{\qquad} \times 37) + (1 \times \underline{\qquad}) = 370 + \underline{\qquad} = \underline{\qquad}$

   _____ Property

3. $(51 \times 14) + (49 \times 14) = (\underline{\qquad} + 49) \times 14 = \underline{\qquad} \times 14 = \underline{\qquad}$

   _____ Property

Solve.

4. What multiplication problem does this drawing represent?

   _____

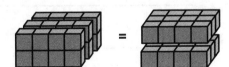

5. Which multiplication property does this drawing represent?

   _____

Name _____

Relating Multiplication and Division

Division splits items into equal groups or parts. Multiplication groups items or parts into one group. You can see the relationship between multiplication and division if you think of items in rows and columns.

Examples

Kit is packing 12 oranges to be sent to the food drive. The oranges are arranged in 3 rows of 4 oranges each.

Kit thought of the problem in this way: The multiplication fact is 3 groups of 4 equals 12. The related division fact is 12 divided by 3 equals 4.

Another student thought of it this way: The multiplication fact is 4 groups of 3 equals 12. The related division fact is 12 divided by 4 equals 3.

This problem has four related facts:

$$3 \times 4 = 12 \qquad 4 \times 3 = 12 \qquad 12 \div 3 = 4 \qquad 12 \div 4 = 3$$

Solve

Sketch each problem. List the related multiplication and division facts.

1. $3 \times 6 =$ ____18____

   $6 \times 3 = 18$ _____

   $18 \div 3 = 6$ _____

   $18 \div 6 = 3$ _____

2. $5 \times 4 =$ _____

   _____

   _____

   _____

3. $2 \times 6 =$ _____

   _____

   _____

   _____

4. $2 \times 1 =$ _____

   _____

   _____

   _____

Name _____

Multiplying by 1-Digit Whole Numbers

There are several strategies you can use to multiply by 1-digit whole numbers.

Examples

A delivery driver makes 19 stops every morning. This morning she picked up 3 packages at each stop. How many packages did she pick up in all?

$$19 \times 3 =$$

Here's one solution:
Break 19 into 10 + 9.
The Distributive Property of Multiplication says that you can write the problem as $(10 \times 3) + (9 \times 3)$.

$$(10 \times 3) + (9 \times 3) = 30 + 27 = 57$$

Here's another solution:
You can solve an easier problem by adding 1 to 19 to make 20.

$$20 \times 3 = 60; 3 \times 1 = 3$$
$$60 - 3 = 57$$

Remember! In a multiplication problem, the answer to a problem is the product.

Solve

Multiply and then describe the strategy you used to find the answer.

1. $16 \times 3 =$ ____48____

   I changed 3 to 6 and 16 to 8. Then I multiplied $6 \times 8 = 48$. _____

2. $21 \times 5 =$ _____

   _____

3. $4 \times 55 =$ _____

   _____

4. 225×3 _____

   _____

5.   31
    $\times$ 3

6.   42
    $\times$ 5

7.   138
    $\times$ 4

8.   21
    $\times$ 8

9.   121
    $\times$ 7

10.   125
    $\times$ 4

11.   138
    $\times$ 8

12.   303
    $\times$ 6

Name _____

Multiplying by 2-Digit Whole Numbers

There are several ways to multiply by a 2-digit whole number.

Examples

David delivers 325 dozen eggs to the local farmers' market. How many eggs are in this delivery?

You can break up 12 into 1 ten and 2 ones.

325×2 ones $= 650$
325×1 ten $= 3250$
$650 + 3250 = 3900$

You can also break up 325 into 300 and 25: $300 \times 12 = 3600$.

Then break up 25 into 5×5: $(5 \times 5) \times 12$ or $(5 \times 12) \times 5 = 60 \times 5 = 300$.

$3600 + 300 = 3900$

Another strategy is to look for patterns.

Double 325 and cut 12 in half: 650×6.

Then double 650 and cut 6 in half: $1300 \times 3 = 3900$.

Solve

Multiply and then describe the strategy you used to solve each problem.

1 $49 \times 50 =$ ___2450___

I added 1 to 49 and multiplied 50 x 50 = 2500. Then I subtracted 50 to find the answer.

2 $43 \times 52 =$ _____

_____

Solve.

3
$$\begin{array}{r} 31 \\ \times\ 23 \\ \hline \end{array}$$

4
$$\begin{array}{r} 62 \\ \times\ 17 \\ \hline \end{array}$$

5
$$\begin{array}{r} 70 \\ \times\ 80 \\ \hline \end{array}$$

6
$$\begin{array}{r} 38 \\ \times\ 19 \\ \hline \end{array}$$

7
$$\begin{array}{r} 60 \\ \times\ 11 \\ \hline \end{array}$$

8
$$\begin{array}{r} 85 \\ \times\ 70 \\ \hline \end{array}$$

9
$$\begin{array}{r} 173 \\ \times\ \ 19 \\ \hline \end{array}$$

10
$$\begin{array}{r} 123 \\ \times\ \ 15 \\ \hline \end{array}$$

11 Toni is buying peaches at the farmers' market. She buys 12 baskets of peaches. Each basket holds 12 peaches. How many peaches does she buy?

_____

Multiplying by Multi-digit Whole Numbers

You can use several strategies to multiply by a multi-digit number. This lesson explains one of them.

Example

Sandra's quilt shop has 12 extra rolls of fabric. Each roll has about 182 meters of fabric. How many meters of fabric are there in all?

Sandra drew an array to work out the multiplication problem 12 × 182.

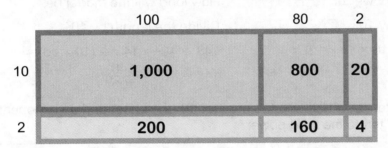

When Sandra added up all the partial products, she found that 12 × 182 = 2184 meters.

Solve

1 149 × 51 = __7599__

2 63 × 22 = _____

3 147 × 17 = _____

4 239 × 41 = _____

5 403 × 39 = _____

6 711 × 15 = _____

7 934 × 21 = _____

8 195 × 39 = _____

9 225 × 25 = _____

10
```
   321
×   23
```

11
```
   602
×   17
```

12
```
   700
×   89
```

13
```
   538
×   49
```

14 About 4 yards of fabric are needed to make one robe for a singer in the Harvest School Glee Club. There are 156 singers in the glee club. How much fabric is needed to make 3 robes for each singer?

_____

Name _____

Multiplying and Dividing by Powers of Ten

You can solve some multiplication problems by using patterns. Look for the patterns that occur when multiplying by powers of ten.

Examples

A micro-organism has a diameter of 0.005 mm. How large will the micro-organism appear if it is viewed under a microscope that has a magnifying power of 10^3?

Multiply the diameter by 10^3.

$0.005 \times 10^3 = 0.005 \times (10 \times 10 \times 10) =$
$0.005 \times 1000 = 5$ mm

Mr. Ruiz plans to make a model of an actual ship that was 148 feet long. The model will be 10^2 times smaller than the actual ship. How long will the model be?

Divide the length by 10^2.

$148 \div 10^2 = 148 \div (10 \times 10) =$
$148 \div 100 = \dfrac{148}{100}$

Remember! A power of ten is written with the number 10 (the base) and an exponent. When you read 10^3, you say, "ten to the third power."

Solve

Multiply or divide by a power of ten.

1 $49 \times 10^3 = 49 \times ($ ___10 × 10 × 10___ $)$

   $49 \times 1000 = $ ___49000___

2 $49 \times 10^2 = 49 \times ($ _____ $)$

   $49 \times$ _____ $=$ _____

3 $525 \div 10^1 = 525 \div$ _____

   $\dfrac{525}{10} = $ _____

4 $525 \div 10^3 = 525 \div$ _____

   $\dfrac{525}{1000} = $ _____

5 $1024 \div 10^3$ _____

6 1776×10^2 _____

7 $225 \div 10^1$ _____

Write the missing word or phrase.

8 When you multiply by a power of ten, the _____ to the right of the 10 tells you how many times to multiply by 10. As the number becomes 10 times greater, the decimal point _____.

9 When you divide by a _____ of ten, the exponent to the right of the 10 tells how many times you divide _____. As the number becomes 10 times _____, the decimal point _____.

Name _____

Division

There are several strategies you can use to solve division problems. You can use an equation, an array, or a model.

Examples

The Button Factory sells boxes of white shirt buttons. Each box holds 9 buttons. There are 162 buttons in a crate. How many boxes of buttons are in each crate?

Write an equation:
Write the problem as 162 ÷ 9 =
Or rewrite the division as multiplication: 9 × _____ = 162.
The solution is 162 ÷ 9 = 18.
There are 18 boxes in each crate.

Use an array:
Draw an array to represent 162.
The array has 9 rows.
Each row has 18 dots.
The solution is 162 ÷ 9 = 18.
There are 18 boxes in each crate.

Draw an area model:
The area is 162. Write the area as 90 + 72.
Use 9 units as the height of the model.
Calculate the width of the model,
10 + 8 = 18.
There are 18 boxes in each crate.

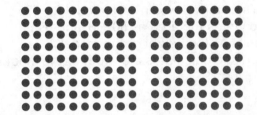

	10	8
9	90 = 9 × 10	72 = 9 × 8

Solve

Find the answer, and tell what strategy you used.

1 147 ÷ 7 ____21____

Strategy: _____

2 156 ÷ 3 _____

Strategy: _____

3 207 ÷ 9 _____

Strategy: _____

4 136 ÷ 8 _____

Strategy: _____

5 119 ÷ 7 _____

Strategy: _____

6 332 ÷ 4 _____

Strategy: _____

Name _____

Dividing by 1-Digit Whole Numbers

Be sure to read a division problem carefully. Try to think of related multiplication facts to use as you read each division problem.

Examples

Cal's tub holds 58 gallons of water. 2 gallons of water pour out of the faucet every minute. How long will it take for the tub to be completely filled?

Cal needs to find 58 gallons ÷ 2 gallons/minute. He breaks 58 into 50 and 8. There are 25 "2s" in 50, and 4 "2s" in 8.

25 + 4 = 29.

It will take 29 minutes for the tub to be completely filled.

Cal's sister draws an area model for 58. She writes the area as 50 + 8, and uses 2 units as the height of the model. Then she calculates the width of the model. 25 + 4 = 29.

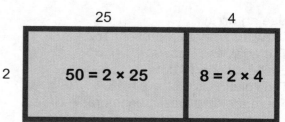

Divide

1. 39 ÷ 3 _____13_____

2. 60 ÷ 5 _____

3. 96 ÷ 4 _____

4. 27 ÷ 9 _____

5. 69 ÷ 3 _____

6. 84 ÷ 6 _____

7. 18 ÷ 6 _____

8. 18 ÷ 3 _____

9. 57 ÷ 3 _____

10. 42 ÷ 7 _____

11. 56 ÷ 4 _____

12. 56 ÷ 8 _____

13. Explain the strategy you used to find the answer to exercise 12.

_____

_____

Dividing by 1-Digit Whole Numbers with a Remainder

Numbers do not always divide into even groups. Sometimes there are numbers left over. Be sure to read each problem carefully. Some problems may ask you to find the number of groups, others may ask you to find the number in each group.

Examples

Dale is placing 63 photos in a photo album. Each page of the album holds 6 photos. How many full pages will Dale have when he is finished?

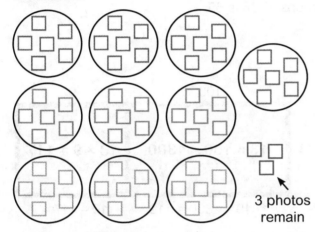

3 photos remain

Dale knows he has to divide 63 by 6. He also knows that $60 \div 6 = 10$. So he'll have 10 full pages and 3 photos left over.

Dale writes his solution as $63 \div 6 = 10$ R3.

Juan has 74 photos and 9 pages in his photo album. How many photos will be on each page?

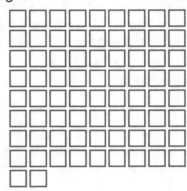

Juan writes the problem as $74 \div 9$. He knows the answer has to be less than 9 because $9 \times 9 = 81$.

$72 \div 9 = 8$. So he'll have 8 photos on 9 full pages and 2 remaining photos.

Juan writes his solution as $74 \div 9 = 8$ R2.

Remember! The remainder in a division problem is written *remainder* (number) or just R (number).

Divide

1. $43 \div 3$ ___14 R1___

2. $64 \div 5$ _____

3. $66 \div 4$ _____

4. $47 \div 9$ _____

5. $24 \div 5$ _____

6. $84 \div 9$ _____

7. $18 \div 7$ _____

8. $28 \div 5$ _____

9. $58 \div 6$ _____

10. $42 \div 8$ _____

11. $83 \div 4$ _____

12. $49 \div 6$ _____

Name _____

Finding Quotients

You can use place value, models, or the properties of multiplication and addition to help you solve division problems.

Examples

1,419 students signed up for Field Day activities. The students formed teams of 13 players for the competition. How many teams were formed?

Define the problem as 1419 ÷ 13.

There are 100 "13s" in 1419, and 1419 − 1300 = 119.
10 groups of 13s is 130. That's too big, but 5 groups of 13 is 65.
119 − 65 = 54. 4 groups of "13s" is 52.

There are 2 students left over.
1419 ÷ 13 = 109 R2.

You can also draw an area model of 1419

One side of the area is 13 units, so use 13 as the height. Now find the width of the rectangle if the area is 1300 + 117 and the height is 13.

1419 ÷ 13 = 109 R2.

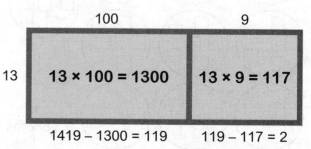

Remember! In a division problem, the *dividend* is the number to be divided, the *divisor* is the number by which the dividend will be divided, and the *quotient* is the result of dividing one number by another.

Divide

1. 399 ÷ 35 ___11 R14___

2. 630 ÷ 52 _____

3. 496 ÷ 44 _____

4. 883 ÷ 28 _____

5. 794 ÷ 62 _____

6. 1496 ÷ 32 _____

7. 7564 ÷ 93 _____

8. 1024 ÷ 80 _____

9. 1956 ÷ 42 _____

10. Ted worked a problem where the divisor was 15 and the quotient included a remainder of 17. What do you know about the remainder? Explain.

_____

Problem Solving

You can use a guess-and-check strategy to solve some problems.

Example

Cindy raises chickens, goats, and ducks. She has 24 animals in all. There are 3 times more goats than ducks. The number of chickens can be divided by 2 and 3.

First, list what you know:
- There are 24 animals in all.
- The number of goats is 3 times the number of ducks.
- The number of chickens is divisible by 2 and 3.

Plan: You can use a guess-and-check strategy to help solve the problem. A chart might help you keep track of the information.

Solve: Organize your guesses in a table.

Write your solution: Cindy has 3 ducks, 9 goats, and 12 chickens.

Ducks (D)	Goats (G)	Chickens (C)	Correct?
	$3 \times D = G$	$C = 24 - (D + G)$	
1	3	20	No, 20 not divisible by 3
2	6	16	No
3	9	12	Yes
4	12	8	No
5	15	4	No
6	18	0	No, she has 3 kinds of animals

Solve

1. Paula collects postage stamps of animals. She has 4 times as many stamps that show tigers as stamps that show leopards. The number of stamps with elephant pictures is divisible by 7. Paula has 48 stamps in all. How many elephant stamps does she have? Create a chart to help find the answer.

   _____

2. Scotty saw a pattern when he divided by 5: Any dividend with a 0 or 5 in the ones place can be divided by 5 with no remainder. What can you say about the quotient if the divisor in a problem is 5 and the dividend has a 1 or a 6 in the ones place?

   _____

   _____

Chapter 3 Test

Name _____

Simplify. Find the product. Name the property.

1 $12 \times 37 = ($ _____ $\times 37) + (2 \times$ _____ $) =$ _____

_____ Property

2 $(51 \times 19) + (49 \times 19) = ($ _____ $+ 49) \times 19 =$ _____ $\times 19 =$ _____

_____ Property

Find the product. List the related multiplication and division facts.

3 $5 \times 8 =$ _____   _____   _____   _____

4 $7 \times 9 =$ _____   _____   _____   _____

Multiply.

5 114
× 4

6 90
× 80

7 44
× 11

8 402
× 13

9 10.6×10^2 _____

10 525×10^2 _____

11 33×10^3 _____

Divide.

12 $101 \div 10^1$ _____

13 $747 \div 10^2$ _____

14 $506 \div 10^3$ _____

Solve the following word problems.

15 Aaron and Kate cut a rope into 8-foot lengths. How many lengths can they get from 78 feet of rope?

_____

16 A camp prepares 67 life jackets for rafts that will travel down the river. Each raft must have 9 life jackets. How many rafts can they equip? _____

How many extra life jackets will be available? _____

17 An expedition to study tropical plants needs to bring 1,525 pounds of food, tents, and other supplies for a 15-day trip. Ten mules will carry all the supplies. How much will each mule carry if each carries about the same weight? _____

18 Researchers need to ship 4,500 pounds of dinosaur fossils back to the research lab. The fossils will be packed in boxes that hold 100 pounds. It costs $75.00 to ship each container. How much will it cost to ship the fossils? _____

19 Ronny took dozens of photographs during his trip to Glacier National Park. He has 4 times as many moose photos as brown bear photos. The number of eagle photos is divisible by 6. Ronny has 50 photos in all. How many moose photos does he have? Draw a table to help you find the answer.

_____

DRAWING OF TABLE:

20 The school lunchroom staff makes 40 sandwiches for the field trip. They make three times as many cheese sandwiches as chicken sandwiches. The rest of the sandwiches are made with hazelnut spread. If they make 15 cheese sandwiches, how many are made with hazelnut spread? Draw a table to help you find the answer.

_____

DRAWING OF TABLE:

21 Gary has $40.00 to spend on camping supplies. He wants to buy a water bottle, compass, insulated lunch bag, and mirror. Does he have enough money to also buy a thermal blanket? Explain.

_____

_____

_____

Campers' Corner	
Blister cream	$7.50
Compass	$8.50
Insulated lunch bag	$4.00
Water bottle	$17.00
Mirror	$1.50
Thermal blanket	$12.50

Models for Multiplying Decimals

There are several ways to show the multiplication of decimals.

Examples

If one item costs $0.32, what is the cost of 6 items?

First, estimate what the cost might be.
3 × 33 is a little less than 100, so 6 × 33 is a little less than 200.

You can estimate that the cost will be a little less than $2.00.

You can use a hundredths grid to find the correct answer.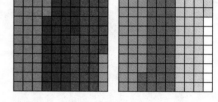
You have 19 whole columns shaded and 2 individual
squares shaded. Nineteen columns of tenths equal 1.90;
2 individual squares equal 2 hundredths, or 0.02.

So 0.32 × 6 = 1.90 + 0.02 = 1.92.

The solution is 6 × $0.32 = $1.92.

You can also use place value to find your answer.
Break 0.32 into 3 tenths and 2 hundredths.
6 × (3 tenths + 2 hundredths) = 6 × (0.3 + 0.02) = (6 × 0.3) + (6 × 0.02) = 1.8 + 0.12 = 1.92.

The solution is 6 × $0.32 = $1.92.

Solve

Estimate. Then multiply and describe your strategy.

1 Estimate: 0.21 × 4 = _____.80_____

Multiply: 0.21 × 4 = _____

Strategy: _____

2 Estimate: 0.11 × 3 = _____

Multiply: 0.11 × 3 = _____

Strategy: _____

3 Estimate: 0.25 × 4 = _____

Multiply: 0.25 × 4 = _____

Strategy: _____

4 Estimate: 0.51 × 6 = _____

Multiply: 0.51 × 6 = _____

Strategy: _____

5 Estimate: 0.96 × 4 = _____

Multiply: 0.96 × 4 = _____

Strategy: _____

6 Estimate: 0.80 × 5 = _____

Multiply: 0.80 × 5 = _____

Strategy: _____

Name _____

Multiplying Decimals

You can use models to help you multiply decimals. But be sure to estimate first.

Example

At the state fair, dried apple slices are $0.39 a bag. What is the cost of 7 bags?

First, estimate the product. The answer will be a little less than 7 × 0.40, or 2.80. Now, use hundredths squares to help find the answer.

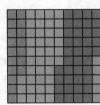

Count 27 shaded columns, which is (20 + 7) × 10.

10 columns equal 1 whole. So 27 columns represent 2 wholes and 7 tenths, written 2.7. There are 3 units. This represents 3 hundredths, written 0.03.

2.7 + 0.03 = 2.73.

7 bags of apple slices cost $2.73.

Solve

Estimate and multiply to complete the table.

		Estimate	Product
❶	2.1 × 2 =	> 4	4.2
❷	3.93 × 3 =		
❸	2.22 × 8 =		
❹	8.24 × 4 =		
❺	4.01 × 7 =		

		Estimate	Product
❻	4.12 × 3 =		
❼	5.54 × 3 =		
❽	0.98 × 7 =		
❾	7.03 × 7 =		
❿	1.56 × 4 =		

⓫ Small jars of prize-winning jam cost $1.39 each. What is the cost of 6 jars of jam?

_____

Name _____

Multiplying Decimals by Powers of Ten

You can multiply decimals by powers of ten.

Example

A marker costs $0.39. What is the cost of ten? What is the cost of a thousand?

Remember! A power of ten is written with the digit 10 (the base) and an exponent. When you read "0.02×10^3," you say "2 hundredths times 10 to the third power."

For ten markers, write 0.39×10 and estimate the cost at about $4.

Then multiply:
$0.39 \times 10 = 0.3 \times 10 + 0.09 \times 10$
$3.0 + 0.9 = 3.90$

10 markers cost $3.90.

When multiplied by 10, the decimal point moved one place to the right.

For a thousand markers, write 0.39×10^3 and estimate the cost at about $400.

Then multiply:
$0.39 \times 10^3 = 0.39 \times (10 \times 10 \times 10)$
$0.39 \times (1000) = 0.3 \times (1000) + 0.09 \times (1000)$
$300 + 90 = 390$

10^3 markers cost $0.39 \times 10^3 = \$390$.

When multiplied by 10 three times, the decimal point moved 3 places to the right.

There is a pattern. Every time you multiply by 10, you move the decimal point one place to the right.

Solve

Estimate. Then multiply.

1 $17.76 \times 10^2 =$

Estimate: _____ < 1,800 _____

Product: _____

3 $1.10 \times 10^4 =$

Estimate: _____

Product: _____

2 $10.24 \times 10^3 =$

Estimate: _____

Product: _____

4 $9.9 \times 10^5 =$

Estimate: _____

Product: _____

Choose from the following terms to complete the sentences below:

product	right	exponent

5 The _____ above the 10 tells how many times you multiply by 10. The

_____ becomes ten times greater every time the exponent increases by 1. Every

time you multiply by 10, the decimal point moves one place to the _____ .

Strategies for Multiplying Decimals

There is more than one way to multiply decimals.

Examples

Suzanne's planter box is 0.6 square yards. She wants to plant tomatoes in 0.3 of the box. How large is the tomato section?

One way to solve is to multiply the decimals just as you would whole numbers.

$$\begin{array}{r} 0.6 \\ \times\ 0.3 \\ \hline 18 \end{array}$$

Then count the number of decimal places in the factors.
That is how many decimal places the product should have.

$$\begin{array}{r} 0.6 \quad \longleftarrow 1 \text{ decimal place} \\ \times\ 0.3 \quad \longleftarrow 1 \text{ decimal place} \\ \hline 0.18 \quad \longleftarrow 2 \text{ decimal places} \end{array}$$

The tomato section is 0.18 square yards.

Another way to solve is by using a hundredths grid.

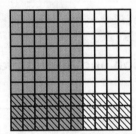

6 tenths of the grid is shaded. This stands for the planter box. 3 tenths of the same grid is striped. This stands for the tomato section. 18 hundredths are both striped and shaded. This stands for the product of 0.6 and 0.3.

The tomato section is 0.18 square yards.

Solve

Estimate and multiply to complete the table. You can use the grids to help.

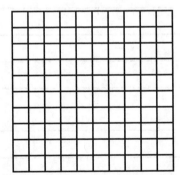
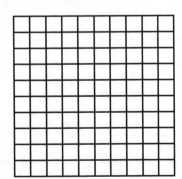
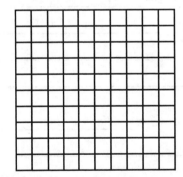

		Estimate	Product
1	0.2 × 0.2 =	0.04	0.04
2	0.4 × 0.7 =		
3	0.3 × 0.8 =		

		Estimate	Product
4	0.6 × 0.5 =		
5	0.9 × 0.8 =		
6	0.2 × 0.02 =		

Name _____

Relating Multiplication and Division of Decimals

You can use models or multiplication to divide decimals. Be sure to estimate first.

Examples

Roy has 2 pounds of rice. He puts the rice in containers that each hold 0.40 pounds. How many containers does he use?

.40 × 10 = 4, so you know that .40 × 5 will be 2.

Roy will use 5 containers.

Knowing multiplication facts can help you with decimal division. For every multiplication fact, there will be three other related facts.

Multiplication fact: 5 × 0.4 = 2.
Related facts:
2 ÷ 0.4 = 5.
0.4 × 5 = 2
2 ÷ 5 = 0.4

Solve

Solve each problem. Then list the related decimal multiplication and division facts.

1 2.4 ÷ 8 = _____ 0.3 _____

_____ 2.4 ÷ 0.3 = 8 _____

_____ 0.3 × 8 = 2.4 _____

_____ 8 × 0.3 = 2.4 _____

4 2 ÷ 0.1 = _____

_____

_____

_____

2 6 ÷ 0.2 = _____

_____

_____

_____

5 8 ÷ 0.4 = _____

_____

_____

_____

3 2.2 ÷ 10 = _____

_____

_____

_____

6 1.6 ÷ 5 = _____

_____

_____

_____

7 When a whole number is divided by a decimal, is the quotient greater than or less than the dividend? _____

Dividing Decimals

Use any strategy to help you divide decimals. But be sure to estimate first.

Examples

A relay race is 5.68 miles long. Each team has 4 runners, and each member runs the same distance. What distance does each team member run?

Estimate that each team member must run between 1 and 2 miles. 1 mile for each runner falls short of the 5.68 miles; 2 miles each is too great a distance.

Break 5.68 into 4 + 1 + 0.40 + 0.28.

Divide each number by 4 and add: 1 + .25 + .1 + .07 = 1.42

You can also use 6 grids to represent the 5.68 miles and divide the 5 whole grids and the 68 squares into 4 equal parts.

| Runner 1 | Runner 2 | Runner 3 | Runner 4 |

Each team member ran 1.42 miles.

Solve

Estimate and divide to complete the table.

		Estimate	Quotient
1	1.2 ÷ 6	0.2	0.2
2	4.4 ÷ 2		
3	5.6 ÷ 8		
4	56 ÷ 0.8		
5	3.9 ÷ 3		
6	6.4 ÷ 4		
7	35 ÷ 0.7		

Name _____

Dividing Decimals by Powers of Ten

You can use a pattern to divide decimals by powers of ten.

Example

$550.5 \div 10^3$

Divide:

$550.5 \div 10^3 =$
$550.5 \div (10 \times 10 \times 10) =$
$550.5 \div 1,000 =$
$(500 + 50 + 0.5) \div 1,000 =$
$(500 \div 1,000) + (50 \div 1,000) + (0.5 \div 1,000) =$
$0.5 + 0.05 + 0.0005 = 0.5505$
So, $550.5 \div 10^3 = 0.5505$

Notice the pattern. Everytime you divide by 10, you move the decimal point one place to the left.

Remember! A power of ten is written with the digit 10 (the base) and an exponent. When you read "$0.03 \div 10^4$," you say "3 hundredths divided by 10 to the fourth power."

Solve

Write the missing numbers.

1. $1.5 \div 10^3 =$

   $1.5 \div (10 \times 10 \times 10) =$

   $1 \div (10 \times \underline{\quad 10 \quad} \times 10) + 0.5 \div (\underline{\quad 10 \quad} \times 10 \times 10) =$

   $(\underline{\quad 0.001 \quad}) + (0.0005) = \underline{\quad 0.0015 \quad}$

2. $0.9 \div 10^2 =$

   $0.9 \div (\underline{\qquad\qquad} \times \underline{\qquad\qquad}) =$

   $0.9 \div (100) = \underline{\qquad\qquad}$

Estimate. Then divide.

3. $17.76 \div 10^2 =$ Estimate: _____

   Quotient: _____

4. $10.24 \div 10^3 =$ Estimate: _____

   Quotient: _____

Name _____

Strategies for Dividing Decimals

You can use the relationship between multiplication and division to divide decimals.

Example

Every year the local bank pays interest on savings accounts. The interest is 0.03 of the balance. Donnie's savings earned $3.60. How much money is in Donnie's savings account?

First, Donnie writes an equation: 0.03 × [balance] = $3.60. Because multiplication and division are related, Donnie rewrites the equation as [balance] = $3.60 ÷ 0.03. He reasons that the answer will include the number 12 because 36 ÷ 3 = 12. He also knows the answer will have to be greater than $3.60, so he estimates the balance to be $120.

Donnie decides to solve a simpler problem: 12 × 3 = 36.
This is related to 120 × 3 = 360 and 360 ÷ 3 = 120.

Then Donnie writes:

$$
\begin{array}{rcl}
360 \div 3 &=& 120 \\
360 \div 0.3 &=& 1{,}200 \\
360 \div 0.03 &=& 12{,}000 \\
36.0 \div 0.03 &=& 1{,}200 \\
3.60 \div 0.03 &=& 120
\end{array}
$$

There is $120 in Donnie's savings account.

Solve

Estimate and divide to complete the table. You can use the grids to help.

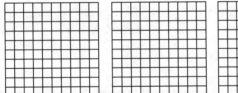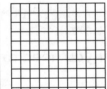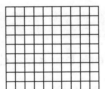

		Estimate	Quotient
1	120 ÷ 0.6 =	200	200
2	44 ÷ 0.02 =		
3	56.7 ÷ 7 =		

		Estimate	Quotient
4	567 ÷ 0.7 =		
5	399 ÷ 0.3 =		
6	0.64 ÷ 4 =		

7 Barbara owes $9,000 on her car loan. She pays interest every month on the amount that is still owed. The interest is 0.01% per month. Write a division or multiplication equation to find the interest on the car loan this month.

_____

How much is the interest this month? _____

Problem Solving

You can work backward to solve some problems.

Example

Darla received her allowance on Saturday evening. The following Monday, she spent $2.50 on a hayride. On Wednesday, she earned $3.00 milking the cows. By Friday, Darla had $8.00 left. How much was her allowance?

First, list what you know.
- She spent $2.50.
- She earned $3.00.
- By Friday, she had $8.00

Write an equation.

Saturday	Monday	Wednesday	Friday
[allowance]	− $2.50	+ $3.00	= $8.00

Start at the end and work backward.

Saturday	Monday	Wednesday	Friday
$7.50	+ $2.50	− $3.00	= $8.00

Check to be certain that your solution makes sense: $7.50 − $2.50 + $3.00 = $8.00.

Darla's allowance is $7.50

Solve

Solve each problem. Work backward if you need to.

1. June picks several baskets of apples. She keeps half the baskets for herself. She gives the rest to 3 friends. Each friend got 2 baskets of apples. How many baskets of apples did June pick?

   12 baskets; 2 × 3 = 6 baskets; 6 baskets × 2 = 12 baskets
   _____

2. Myrna and Barry have a goat farm. They began with a certain number of goats, and every year after that, the goat population doubled. In 5 years, Myrna and Barry had 144 goats. How many goats did they begin with?

   _____

   _____

3. Marvin wants to know the age of Mr. Fogel's horse. Mr. Fogel told Marvin that if he added 10 years to the age of the horse and then doubled it, the horse would be 44 years old. How old is the horse?

   _____

   _____

Estimate. Then multiply. You can use the grids to help.

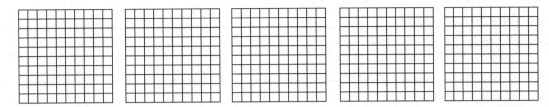

		Estimate	Product
1	1.3 × 2		
2	4.5 × 3		
3	13.2 × 8		
4	4.24 × 3		
5	2.01 × 5		

		Estimate	Product
6	0.3 × 0.3		
7	0.4 × 0.6		
8	0.3 × 0.8		
9	0.6 × 0.8		
10	1.2 × 0.4		

Estimate. Then divide.

		Estimate	Quotient
11	1.2 ÷ 6		
12	4.4 ÷ 2		
13	5.6 ÷ 8		
14	35.7 ÷ 7		
15	3.9 ÷ 3		

		Estimate	Quotient
16	15 ÷ 0.5		
17	66 ÷ 0.03		
18	56 ÷ 0.8		
19	284 ÷ 0.4		
20	396 ÷ 0.33		

Estimate. Then multiply or divide.

21 10.66×10^3 Estimate: _____

Product: _____

22 11.95×10^2 Estimate: _____

Product: _____

23 14.92×10^1 Estimate: _____

Product: _____

24 $81.52 \div 10^0$ Estimate: _____

Quotient: _____

25 $19.99 \div 10^3$ Estimate: _____

Quotient: _____

26 $11.55 \div 10^2$ Estimate: _____

Quotient: _____

Estimate. Then multiply and describe your strategy.

27 Estimate: $0.7 \times 5 =$ _____

Multiply: $0.7 \times 5 =$ _____

Strategy: _____

Solve. List the related multiplication and division facts.

28 $5 \div 0.1 =$ _____

_____

_____

_____

Solve.

29 The following amounts were added to a number: 7.2×10^2, 4.6×10^2, 2.7×10^2, and 6.2×10^2. After the amounts were added, the total was 34.4×10^2. What was the original number? Work backward to solve.

_____

_____

30 An average adult should drink about 2 liters of water a day. How many 1-liter bottles of water are needed a day for 1,000 adults?

_____

A case of 1-liter bottles of water holds 24 bottles. How many are needed for 1,000 people per day?

_____

31 A table shows a population of 10^4 people for a community. How many people are in the community?

_____

If each person in the community was given 0.3 pounds of rice each day for a week, how much rice was needed?

_____

A pound of rice is 16 ounces. How many ounces of rice are in 0.3 pounds?

_____

Order of Operations

If there are more than two operations in one expression, there are rules to tell you which operations to complete first.

Example

There are 28 passengers on a bus going to the mall. At one stop, 2 groups of 3 passengers leave the bus to go shopping and 3 groups of 4 passengers leave the bus to eat lunch. How many passengers are left on the bus?

This problem can be written as $28 - (2 \times 3) - (3 \times 4)$

To solve a problem correctly, you must always follow the order of operations.

Step 1: Simplify terms within **parentheses**.
Step 2: Simplify terms with **exponents**.
Step 3: **Multiply** and **divide** from left to right.
Step 4: **Add** and **subtract** from left to right.

$28 - (2 \times 3) - (3 \times 4) =$
$28 - 6 - 12 =$
$22 - 12 = 10$

There are 10 passengers remaining on the bus.

Solve

Follow the directions to simplify each expression.

1 $(15 - 5)^3 - (40 \times 5) \times (8 - 5) + 10^2 =$

Simplify inside parentheses: $\underline{(10)^3 - (200) \times (3) + 10^2}$ _____

Simplify exponents: _____

Multiply: _____

Add and subtract from left to right: _____

Final answer: _____

2 $(25 \times 4) \times (9 - 4) + (10 - 5)^2 - 10^1 =$

Simplify inside parentheses: _____

Simplify exponents: _____

Multiply: _____

Add and subtract from left to right: _____

Final answer: _____

Expressions with Parentheses

Simplifying terms within parentheses is the first step in the order of operations.

Example

Ms. Chou's 5th grade class is working to explain the steps needed to simplify the following expression: $5^2 + (9 \times 3) - 4$

The first step is to simplify the terms within parentheses. $5^2 + 27 - 4$

Then simplify terms with exponents: $25 + 27 - 4$

Finally add and subtract from left to right: $52 - 4 = 48$

One way to remember the order of operations is by using a memory device:

Pink Elephants Dislike Mice And Snails
(Parentheses Exponents Divide Multiply Add Subtract)

Solve

Simplify and solve. Show your work.

1 $(20 + 7) - 3$

$\underline{27 - 3 = 24}$

2 $8 + (5 \times 8) \times 6$

_____

3 $(5 \times 2) \div 2^2$

_____

4 $(43 - 8) - (22 - 18)^2$

_____

5 $3^2 \times (2 \times 5) + 7 - (2 \times 3)$

_____

6 $4 + 2^2 \times (5 \times 2) \div 4$

_____

7 $(40 \div 5) + (22 - 12)^2$

_____

8 Use parentheses to show the operations used: $6 \times 4 + 12 - 8 + 2^2 = 32$

_____

_____

Expressions with Parentheses, Brackets, and Braces

Some math expressions contain brackets [ ] and braces { }, in addition to parentheses. This lesson will help you learn how to evaluate these expressions.

Example

The order for parentheses, brackets, and braces is

1. First parentheses ( ) $3\{4[11 + 4(400 + 100) + 289]\} =$

$3\{4[11 + 4(500) + 289]\} = 3\{4[11 + 2000 + 289]\}$

2. Then brackets [ ] $3\{4[11 + 2000 + 289]\} =$

$3\{4[2300]\} =$

3. Finally, braces { } $3\{4[2300]\} =$

$3\{9200\} = 27600$

Remember! Remember the order of operations:
Parentheses, **Exponents**, **Multiply** and **Divide** from left to right,
Add and **Subtract** from left to right.

Solve

Simplify each expression.

1 $\{[3(15 + 5) + 27] \times 4\}$

Simplify inside the parentheses: $\{[3(20) + 27] \times 4\}$ _____

Simplify inside the brackets: _____

Simplify inside the braces: _____

$\{[3(15 + 5) + 27] \times 4\} =$ _____

2 $3\{[2(3 + 6) + 9] - 24\}$

Simplify inside the parentheses: _____

Simplify inside the brackets: _____

Simplify inside the braces: _____

$3\{[2(3 + 6) + 9] - 24\} =$ _____

3 $[(10 \times 2 + 8) \div 14]^2$

Simplify inside the parentheses: _____

Simplify inside the brackets: _____

$[(10 \times 2 + 8) \div 14]^2 =$ _____

Interpreting Numerical Expressions

Knowing how to simplify numerical expressions is a useful tool to have. However, you also need to be able to describe the meaning of the expression without having to do the calculations.

Examples

A word problem instructs a student to "double 4 and then add 32."
This expression can be written as $2 \times 4 + 32$.

A second word problem asks the reader to explain how the expression $3 \times (10^6)$ relates to the expression 10^6.

We know that $3 \times (10^6)$ is 3 times larger than 10^6, because $3 \times (10^6)$ means there are three groups of 10^6.

Solve

Write an expression for each group of steps. Remember the Order of Operations.

1 Multiply 3 times 2 squared and then subtract the product from 12.

$\underline{12 - (3 \times 2^2)}$ _____

2 Divide 18 by 2 and then subtract the quotient from 22.

_____

3 Multiply the difference between 5 and 2 by 3 squared.

_____

4 Find the product of 6 and 9, then add 3.

_____

Write a description for each expression.

5 $(4 + 2)^2 \times (5 \times 2)$

_____

6 $6 \times (5 \times 3) - 50$

_____

7 $3^2 \times 2^2 + (5 \times 4)$

_____

Write < or >. Show how to simplify the expressions.

8 $(45 \div 9) \times 7$ _____ $24 \times 3 - 30$

_____

Problem Solving
Patterns can be found everywhere. Looking for patterns in data can help you solve problems.

Example

Aunt Linda has a plan for giving her nieces and nephews a gift of money. Dan is 10 and gets $10. Ron is 12 and gets $14, Emily is 14 and gets $18, and Sheryl is 16 and gets $22. Her oldest niece, Katie, is 18. How much money does Katie get?

First, list what you know: A table will help you keep track of the data.

Plan: Look for a pattern in the data. The gifts appear to be related to each child's age.

Solve: Look through the data in the chart. See if you can find a pattern. It appears as if the allowance has a base of $10 + $[2 × (age − 10)].

Write your solution: Katie gets $26.00

Look Back: Check to be certain that your solution makes sense. Does your answer fit the pattern?

Age	Allowance	
10	$10	10 + (0 × 2) = 10
12	$14	10 + (2 × 2) = 14
14	$18	10 + (4 × 2) = 18
16	$22	10 + (6 × 2) = 22
18		10 + (8 × 2) = 26

Solve

1. Scott writes the following series of numbers: 0.007, 0.027, _0.047_, 0.067, 0.087

   What is the missing number? _The difference between numbers is 0.020; 0.027 + 0.020 = 0.047_

2. Kip is at an auction for antique motor bikes. The opening bid is $2,000. The next 3 bids are $2,500, $3,000, and $3,500. The bike sells on the eighth bid. If the pattern continues, what is the final, winning bid? Create a table.

   _____

   DRAWING OF TABLE

3. Dylan has 8 U.S. coins. The coins have a total value of $0.56.

   What are the eight coins? _____

Name _____

Follow the directions to simplify each expression.

1 $(15 - 5)^3 - (20 \times 5) \times (8 - 4) + 10^2 =$

Simplify inside parentheses: _____

_____

Simplify exponents: _____

Multiply: _____

Add and subtract from left to right: _____

_____

Final answer: _____

2 $(30 \times 5) \times (6 - 2) + (10 - 5)^2 - 10^1 =$

Simplify inside parentheses: _____

_____

Simplify exponents: _____

Multiply: _____

Add and subtract from left to right: _____

_____

Final answer: _____

Use the order of operations to simplify and solve. Show your work.

3 $(7 \times 3) \div 3^2$

_____

4 $(52 + 9) - 14$

_____

5 $12 + (4 \times 5) \times 8$

_____

6 $5 + 2^2 \times (5 \times 2) \div 5$

_____

7 $(60 \div 10) + (12 - 8)^2$

_____

8 $2^2 \times 3^2 + 6 - (8 \times 4)$

_____

Simplify each expression.

9 $\{[3(5 + 5) + 25] \times 2\}$

Simplify inside the parentheses: _____

Simplify inside the brackets: _____

Simplify inside the braces: _____

$\{[3(5 + 5) + 25] \times 2\} =$ _____

10 $3\{[2(4 + 5) + 8] - 20\}$

Simplify inside the parentheses: _____

Simplify inside the brackets: _____

Simplify inside the braces: _____

$3\{[2(4 + 5) + 8] - 20\} =$ _____

11 $[(10 \times 3 + 4) \div 17]^2$

Simplify inside the parentheses: _____

Simplify inside the brackets: _____

$[(10 \times 3 + 4) \div 17]^2 =$ _____

Write an expression for each description.

12 Multiply 4 times 2 squared, and then subtract the product from 15.

_____

13 Divide 36 by 6, and then subtract the quotient from 54.

_____

14 Multiply 7 squared by the difference between 8 and 6.

_____

15 Find the product of 12 and 15, and then add 3

_____

Write a description for each expression.

16 $(4 + 2)^2 - 16$

_____

17 $8 \times (6 \times 7) - 50$

_____

18 $48 - 3^2 \times 2^2$

_____

19 $7 \times (9 \times 2) - 2^2$

_____

Solve.

20 $(150 - 25) - (25 - 5^2)$

_____

21 $\{[4 \times (17 + 8) + 27] \times 4\}$

_____

Solve.

22 Elizbeth needs to find the missing number in the following series.

0.459, 0.859, _____, 1.659, 2.059

What is the missing number?

_____

23 Twins Don and Dan are working together to solve a math problem. Don finds the answer and tells Dan: "Square the sum of 6 plus 2, and then subtract 30." Write the numerical expression that matches Don's directions.

_____

24 A gallon of fat-free milk costs $3.99. A year ago, the gallon of milk cost 3 percent (0.03) less. How much did a gallon of milk cost a year ago?

_____

25 Devon works a total of 21 hours on Friday, Saturday, and Sunday. He works twice as many hours on Sunday as he does on Saturday. He works 6 hours on Friday. How many hours does he work on Saturday?

_____

Name _____

Finding Least Common Multiples

Fractions can have like denominators (for example, $\frac{3}{4}$ and $\frac{1}{4}$).

Fractions can also have unlike denominators (for example, $\frac{1}{3}$ and $\frac{1}{4}$). You can rewrite fractions with unlike denominators to have like denominators. Finding the least common multiple (LCM) will help you to do this.

Example

The lease common multiple (LCM) of two numbers is the smallest number (not zero) that is a multiple of both numbers. What is the LCM of 3 and 4?

Write the multiples of 3:	⓪	3	6	9	⑫	15	18	21	㉔	. . .
Then write the multiples of 4:	⓪	4	8	⑫	16	20	㉔	28	32	. . .

Circle the multiples that are common to both 3 and 4. Disregard the zeros. You can then see that the least common multiple (LCM) of 3 and 4 is 12.

Solve

Write the multiples for each pair of numbers. Identify the LCM for each pair.

1 4, 5

Multiples: _4, 8, 12, 16, 20, 24, 28_____

Multiples: _5, 10, 15, 20, 25, 30_____

LCM of 4, 5: _20_____

2 8, 6

Multiples: _____

Multiples: _____

LCM of 8, 6: _____

3 4, 6

Multiples: _____

Multiples: _____

LCM of 4, 6: _____

4 5, 6

Multiples: _____

Multiples: _____

LCM of 5, 6: _____

Name _____

Finding Equivalent Fractions

Sometimes to add or subtract fractions, you need to use equivalent fractions. An equivalent fraction is a fraction that has the same value or represents the same part of an object as the original fraction.

Examples

What fraction has a denominator of 8 and is equivalent to $\frac{1}{2}$?

You can draw models to solve the problem.

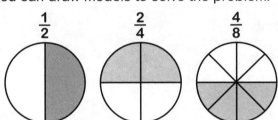

$$\frac{1}{2} \qquad \frac{2}{4} \qquad \frac{4}{8}$$

You can also use multiplication to find equivalent fractions.

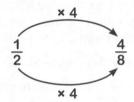

You need a denominator of 8. You know that $2 \times 4 = 8$ and $\frac{4}{4} = 1$. Multiply the numerator and denominator of $\frac{1}{2}$ by 4.

$$\frac{1}{2} \times 1 = \frac{1}{2} \text{ and } \frac{1}{2} \times \frac{4}{4} = \frac{4}{8}.$$

$\frac{4}{8}$ and $\frac{1}{2}$ are equivalent fractions.

Solve

Find the equivalent fraction.

1. equivalent to $\frac{1}{2}$, denominator 16 _____ $\frac{8}{16}$ _____

2. equivalent to $\frac{1}{3}$, denominator 12 _____

3. equivalent to $\frac{1}{5}$, denominator 20 _____

4. equivalent to $\frac{2}{3}$, denominator 15 _____

5. equivalent to $\frac{3}{4}$, denominator 12 _____

6. equivalent to $\frac{3}{4}$, denominator 16 _____

7. equivalent to $\frac{3}{5}$, denominator 20 _____

8. equivalent to $\frac{5}{12}$, denominator 24 _____

9. equivalent to $\frac{4}{5}$, denominator 30 _____

10. equivalent to $\frac{5}{7}$, denominator 49 _____

Name _____

Adding Fractions with Like Denominators

You can add fractions with like denominators.

Examples

When you add fractions with the same denominator, you add the numerators and keep the denominators.

$$\begin{array}{r} \frac{1}{8} \\ + \frac{3}{8} \\ \hline \frac{4}{8} \end{array}$$

Sometimes you can simplify a fraction after adding. To simplify this fraction, divide the numerator and denominator by a number that can divide both. In this instance, both the 4 and 8 of $\frac{4}{8}$ can be divided by 2 and 4. Choose to divide by 4. It is the greater of the two numbers.

$$\frac{4 \div 4}{8 \div 4} = \frac{1}{2}$$

Sometimes when you add, you get an improper fraction. An improper fraction is one with a numerator larger than the denominator. To simply an improper fraction, your divide the numerator by the denominator.

$$\begin{array}{r} \frac{7}{9} \\ + \frac{4}{9} \\ \hline \frac{11}{9} \end{array}$$

$$11 \div 9 = 1 \text{ R2} = 1\frac{2}{9}$$

Solve

Add. Give the answer in simplest terms.

1. $\frac{1}{4} + \frac{1}{4}$ _____ $\frac{2}{4} = \frac{1}{2}$ _____

2. $\frac{2}{9} + \frac{5}{9}$ _____

3. $\frac{1}{7} + \frac{3}{7}$ _____

4. $\begin{array}{r} \frac{3}{31} \\ + \frac{4}{31} \\ \hline \end{array}$

5. $\begin{array}{r} \frac{2}{9} \\ + \frac{1}{9} \\ \hline \end{array}$

6. $\begin{array}{r} \frac{7}{39} \\ + \frac{6}{39} \\ \hline \end{array}$

7. $\begin{array}{r} \frac{14}{23} \\ + \frac{19}{23} \\ \hline \end{array}$

8. $\begin{array}{r} \frac{5}{8} \\ + \frac{5}{8} \\ \hline \end{array}$

9. $\begin{array}{r} \frac{2}{15} \\ + \frac{8}{15} \\ \hline \end{array}$

10. $\begin{array}{r} \frac{4}{9} \\ + \frac{2}{9} \\ \hline \end{array}$

11. $\begin{array}{r} \frac{2}{18} \\ + \frac{1}{18} \\ \hline \end{array}$

Name _____

Subtracting Fractions with Like Denominators

You can subtract fractions with like denominators.

Example

Mr. Hudson shortened a pair of brown pants by $\frac{3}{8}$ of an inch. He shortened a pair of blue pants by $\frac{7}{8}$ of an inch. How many more fractions of an inch did he cut off the blue pants than the brown pants?

Plot the fractions on a number line to find the answer.

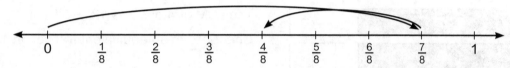

$\frac{7}{8} - \frac{3}{8} = \frac{4}{8}$

Simplify $\frac{4}{8}$ to $\frac{1}{2}$

Mr. Hudson shortened the blue pants $\frac{1}{2}$ inch more.

Solve

Subtract. Give the answer in simplest terms.

1 $\frac{3}{4} - \frac{1}{4}$ _____ $\frac{2}{4} = \frac{1}{2}$ _____

2 $\frac{11}{16} - \frac{3}{16}$ _____

3 $\frac{6}{7} - \frac{3}{7}$ _____

4 $\frac{5}{8} - \frac{3}{8}$ _____

5 $\frac{13}{16} - \frac{7}{16}$ _____

6 $\frac{21}{37} - \frac{4}{37}$ _____

7 $\frac{17}{31}$
$- \frac{7}{31}$

8 $\frac{2}{9}$
$- \frac{1}{9}$

9 $\frac{27}{53}$
$- \frac{6}{53}$

10 $\frac{19}{29}$
$- \frac{14}{29}$

11 $\frac{7}{8}$
$- \frac{5}{8}$

12 $\frac{8}{15}$
$- \frac{1}{15}$

13 $\frac{24}{35}$
$- \frac{13}{35}$

14 $\frac{7}{15}$
$- \frac{2}{15}$

15 $\frac{38}{77}$
$- \frac{25}{77}$

16 $\frac{7}{24}$
$- \frac{5}{24}$

17 $\frac{43}{65}$
$- \frac{38}{65}$

18 $\frac{15}{16}$
$- \frac{11}{16}$

Name _____

Adding Fractions with Unlike Denominators

Models and equivalent fractions can be useful when adding fractions with unlike denominators.

Examples

Kathy makes flavored ice tea by mixing $\frac{3}{4}$ quart of green tea and $\frac{2}{3}$ quart cranberry juice. How much flavored ice tea did Kathy make?

You can use fraction strips to help you add.

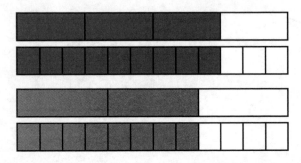

From the fraction strips you can see that

$$\frac{3}{4} = \frac{9}{12} \qquad \frac{2}{3} = \frac{8}{12}$$

$$\text{and } \frac{9}{12} + \frac{8}{12} = \frac{17}{12} = 1\frac{5}{12}$$

Kathy made 1 and $\frac{5}{12}$ quarts of ice tea.

You can also use equivalent fractions to solve this problem.

The LCM of 4 and 3 is 12.

Change $\frac{3}{4}$ to an equivalent fraction in 12ths.

$$\frac{3}{4} \times \frac{3}{3} = \frac{9}{12}$$

Then change $\frac{2}{3}$ to an equivalent fraction in 12ths.

$$\frac{2}{3} \times \frac{4}{4} = \frac{8}{12}$$

Then, add the equivalent fractions:

$$\frac{9}{12} + \frac{8}{12} = \frac{17}{12}$$

Finally, change this improper fraction to a mixed number.

$$\frac{17}{12} = 1\frac{5}{12}$$

Solve

Find equivalent fractions and then add. Change improper fractions to mixed numbers.

① $\frac{1}{4} + \frac{1}{5}$ $\dfrac{5}{20} + \dfrac{4}{20} = \dfrac{9}{20}$

② $\frac{2}{9} + \frac{1}{6}$ _____

③ $\frac{3}{8} + \frac{1}{3}$ _____

Add. Change any improper fractions to mixed numbers.

④ $\frac{1}{2}$
  $+ \frac{3}{7}$
  _____

⑤ $\frac{2}{3}$
  $+ \frac{3}{5}$
  _____

⑥ $\frac{1}{7}$
  $+ \frac{1}{8}$
  _____

⑦ $\frac{1}{7}$
  $+ \frac{1}{6}$
  _____

Name _____

Adding Mixed Numbers with Unlike Denominators

The methods you use to add fractions with unlike denominators are the same you use to add mixed numbers with unlike denominators.

Remember! A mixed number is a number with a whole number part and a fraction part.

Example

In art class, Miranda makes salt dough using $\frac{3}{4}$ cup of flour and $1\frac{1}{3}$ cup salt. What is the total amount of flour and salt she used?

First, find equivalent fractions.

Rewrite $\frac{3}{4}$ as $\frac{9}{12}$ and $1\frac{1}{3}$ as $1\frac{4}{12}$.

You do not need to change the whole number 1 into a fraction.

Then, add $\frac{9}{12} + 1\frac{4}{12} = 1\frac{13}{12}$.

Change the improper fraction $\frac{13}{12}$ to a mixed number:

$\frac{13}{12} = 1\frac{1}{12}$

Finally, add $1 + 1\frac{1}{12} = 2\frac{1}{12}$.

The total amount of flour and salt she used is $2\frac{1}{12}$ cups.

Solve

Find equivalent fractions and add. Change improper fractions to mixed numbers.

1. $1\frac{1}{4} + \frac{1}{6}$ _____ $1\frac{3}{12} + \frac{2}{12} = 1\frac{5}{12}$

4. $2\frac{1}{9} + \frac{5}{6}$ _____

2. $1\frac{2}{3} + \frac{1}{2}$ _____

5. $4\frac{3}{8} + 2\frac{1}{6}$ _____

3. $2\frac{1}{4} + 3\frac{5}{8}$ _____

6. $1\frac{1}{5} + 1\frac{1}{2}$ _____

7. $3\frac{1}{2}$
$+ 1\frac{3}{7}$
_____

8. $2\frac{1}{3}$
$+ 1\frac{3}{5}$
_____

9. $6\frac{5}{7}$
$+ 5\frac{3}{8}$
_____

10. $3\frac{4}{7}$
$+ 4\frac{1}{6}$
_____

Name _____

Subtracting Fractions with Unlike Denominators

The strategies you use to add fractions with unlike denominators can help you subtract fractions with unlike denominators.

Examples

Allen has $\frac{3}{4}$ pound of dog food. He feeds his pet dog, Champ, $\frac{2}{3}$ pound of dog food. How much dog food does Allen have left after feeding Champ?

You can use fraction strips to find how much food is left.

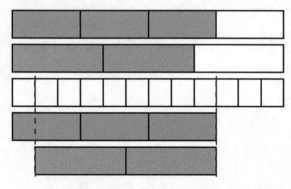

From the fraction strips you can see that

$$\frac{3}{4} = \frac{9}{12} \qquad \frac{2}{3} = \frac{8}{12}$$

$$\text{and } \frac{9}{12} - \frac{8}{12} = \frac{1}{12}$$

There is $\frac{1}{12}$ pound of dog food left.

You can also use equivalent fractions to solve this problem. The LCM of 4 and 3 is 12, so change $\frac{3}{4}$ to an equivalent fraction in 12ths and $\frac{2}{3}$ to an equivalent fraction in 12ths.

$$\frac{3}{4} \times \frac{3}{3} = \frac{9}{12}$$

$$\frac{2}{3} \times \frac{4}{4} = \frac{8}{12}$$

Next, subtract the equivalent fractions:

$$\frac{9}{12} - \frac{8}{12} = \frac{1}{12}$$

Solve

Find equivalent fractions and subtract. Write the difference in simplest form.

1 $\frac{2}{9} - \frac{1}{6}$ _____ $\frac{4}{18} - \frac{3}{18} = \frac{1}{18}$ _____

3 $\frac{3}{8} - \frac{1}{3}$ _____

2 $\frac{1}{2} - \frac{1}{3}$ _____

4 $\frac{3}{4} - \frac{1}{5}$ _____

5 $\begin{array}{r} \frac{1}{2} \\ -\frac{3}{7} \\ \hline \end{array}$

6 $\begin{array}{r} \frac{2}{3} \\ -\frac{1}{4} \\ \hline \end{array}$

7 $\begin{array}{r} \frac{1}{9} \\ -\frac{1}{11} \\ \hline \end{array}$

8 $\begin{array}{r} \frac{1}{6} \\ -\frac{1}{7} \\ \hline \end{array}$

Subtracting Mixed Numbers with Unlike Denominators

The strategies you use to subtract fractions with unlike denominators can help you subtract mixed numbers with unlike denominators.

Example

Ross is making mozzarella cheese at home. He has $4\frac{1}{2}$ teaspoons of salt. He uses $\frac{2}{3}$ teaspoon salt for the mozzarella. How much salt is left when he is finished?

You can see that the fractions have unlike denominators. You need to find equivalent fractions.

Rewrite $4\frac{1}{2}$ as $4\frac{3}{6}$.

Rewrite $\frac{2}{3}$ as $\frac{4}{6}$.

Write the expression $4\frac{3}{6} - \frac{4}{6}$.

You cannot subtract $\frac{4}{6}$ from $\frac{3}{6}$. So you need to rewrite $4\frac{3}{6}$ as $3 + (\frac{6}{6} + \frac{3}{6}) = 3\frac{9}{6}$.

Then rewrite the subtraction expression: $3\frac{9}{6} - \frac{4}{6} = 3\frac{5}{6}$.

Ross will have $3\frac{5}{6}$ teaspoons of salt left.

Solve

Find equivalent fractions and subtract. Simplify improper fractions.

1 $1\frac{1}{4} - \frac{1}{6}$ _____ $1\frac{3}{12} - \frac{2}{12} = 1\frac{1}{12}$ _____

4 $3\frac{3}{8} - 2\frac{1}{6}$ _____

2 $1\frac{2}{3} - \frac{1}{2}$ _____

5 $2\frac{1}{9} - \frac{5}{6}$ _____

3 $1\frac{7}{8} - 1\frac{3}{6}$ _____

6 $3\frac{1}{2} - 2\frac{1}{5}$ _____

7 $\begin{array}{r} 3\frac{1}{2} \\ - 1\frac{3}{7} \\ \hline \end{array}$ **8** $\begin{array}{r} 2\frac{1}{3} \\ - 1\frac{2}{5} \\ \hline \end{array}$ **9** $\begin{array}{r} 5\frac{6}{7} \\ - 4\frac{5}{6} \\ \hline \end{array}$ **10** $\begin{array}{r} 6\frac{7}{8} \\ - 1\frac{2}{3} \\ \hline \end{array}$

Name _____

Reasonable Estimates

When working with fractions, you can use familiar strategies to assess the reasonableness of an estimate.

Examples

Students are designing a mini track so that they can race tiny model cars. One track is $63\frac{1}{8}$ inches long. Another track is $75\frac{3}{4}$ inches long. Estimate the difference in length between these two tracks.

You can use front-end estimation.
Identify the greatest place value in each number.
Subtract. Write zeros for the other place values.

$$75\frac{3}{4}$$
$$- \ 63\frac{1}{8}$$

$$\begin{array}{r} 70 \\ - \ 60 \\ \hline 10 \end{array}$$

The difference is about 10 inches.

You can also use a number line to make estimations.

Estimate the sum of $\frac{5}{6} + \frac{5}{12}$.

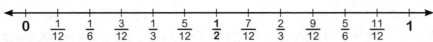

$$0 \quad \frac{1}{12} \quad \frac{1}{6} \quad \frac{3}{12} \quad \frac{1}{3} \quad \frac{5}{12} \quad \frac{1}{2} \quad \frac{7}{12} \quad \frac{2}{3} \quad \frac{9}{12} \quad \frac{5}{6} \quad \frac{11}{12} \quad 1$$

Note that $\frac{5}{6}$ is close to 1 and that $\frac{5}{12}$ is close to $\frac{1}{2}$.

Estimate the sum to be about $1 + \frac{1}{2} = 1\frac{1}{2}$.

Solve

Estimate the sum or difference. Use front-end estimation.

1 $32\frac{2}{9} + 12\frac{1}{7}$ _____ 30 + 10 = 40 _____

3 $77\frac{5}{9} - 21\frac{2}{5}$ _____

2 $45\frac{1}{2} - 33\frac{1}{5}$ _____

4 $34\frac{7}{8} + 22\frac{2}{9}$ _____

5
$$\begin{array}{r} \frac{5}{6} \\ + \ \frac{6}{7} \\ \hline \end{array}$$

6
$$\begin{array}{r} \frac{5}{8} \\ - \ \frac{3}{7} \\ \hline \end{array}$$

7
$$\begin{array}{r} \frac{2}{3} \\ - \ \frac{1}{4} \\ \hline \end{array}$$

8
$$\begin{array}{r} \frac{1}{9} \\ + \ \frac{7}{11} \\ \hline \end{array}$$

Word Problems with Fractions
Some word problems require the addition and subtraction of fractions.

Example

Mr. Goldberg ordered boxes of juice for his class. Each box holds eight cartons and there are 48 cartons of juice in all. One third of the boxes hold orange juice. How many cartons are not orange juice?

You can draw fraction squares to help you solve the problem.

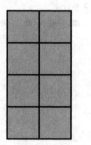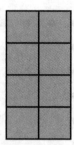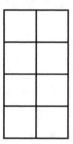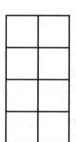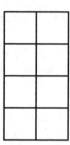

You can see that there are 16 cartons of orange juice. You can then tell that $\frac{2}{3}$ of the 48 cartons, or 32 cartons, do not contain orange juice.

Solve

First write an expression and then solve the problem. Simplify if needed.

1. Mr. Goldberg notices that $\frac{1}{12}$ of the juice boxes hold apple juice and $\frac{1}{16}$ are berry. What fraction of the juice boxes are either apple or berry?

   Expression: $\frac{1}{12} + \frac{1}{16}$ _____

   Solution: $\frac{7}{48}$ are either apple or berry: $\frac{1}{12} + \frac{1}{16} = \frac{4}{48} + \frac{3}{48} = \frac{7}{48}$ _____

2. What fraction of juice boxes holds either orange juice or apple juice?

   Expression: _____

   Solution: _____

3. Explain how to use the number of juice boxes and the given fractions to find the actual number of orange, apple, and berry juice boxes.

   _____

   _____

Name _____

Problem Solving

An effective strategy for problem solving is to draw a diagram as you read a problem. Drawing a diagram can help you "see" the problem. It might even help you see the answer.

Example

Amy wants to fence an area for her dog, Ruby. The area to be fenced is 48 feet long and 36 feet wide. If the fence posts are placed 6 feet apart, how many posts will she need?

Draw a diagram of the area.

Use the diagram and sketch in the location of posts. Make certain there is a post at each corner. The diagram will show that 28 posts are needed.

Check to be certain that your solution makes sense. Altogether there is 168 feet of fence. Each post supports 6 feet of fence: 168 ÷ 6 = 28.

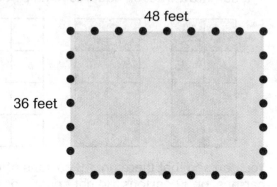

The solution makes sense: 28 fence posts are needed.

Solve

First draw a diagram and then solve.

① Ken and Pete built a bridge made of toothpicks. Ken worked $3\frac{1}{3}$ hours on the bridge, and Pete worked $\frac{1}{2}$ hour more than Ken. How much total time did they spend working on the bridge?

Ken: _____ $3\frac{1}{3}$ _____ hours

Pete: Ken's hours + _____ $\frac{1}{2}$ _____ hours

Total Time = _____

② At Taylor School, 145 fifth-grade students display their projects in the science fair. 15 more girls than boys display their projects. How many girls and how many boys have entered the science fair?

Identify the least common multiple, LCM, for each pair of numbers.

1 5, 6 _____   **2** 7, 8 _____   **3** 3, 4 _____   **4** 7, 6 _____

5 10, 12 _____   **6** 6, 9 _____   **7** 2, 5 _____   **8** 5, 3 _____

Find the equivalent fractions.

9 equivalent to $\frac{1}{2}$, denominator 22 _____

12 equivalent to $\frac{2}{3}$, denominator 12 _____

10 equivalent to $\frac{2}{5}$, denominator 20 _____

13 equivalent to $\frac{3}{5}$, denominator 15 _____

11 equivalent to $\frac{3}{7}$, denominator 35 _____

14 equivalent to $\frac{1}{3}$, denominator 21 _____

Add or subtract. Give answers in simplest terms.

15 $\frac{1}{4} + \frac{1}{4}$ _____

16 $\frac{2}{9} + \frac{5}{9}$ _____

17 $\frac{3}{7} - \frac{2}{7}$ _____

18 $\frac{3}{3} + \frac{2}{5}$ _____

19 $\frac{9}{17} - \frac{5}{17}$ _____

20 $\frac{5}{8} - \frac{3}{8}$ _____

Add or subtract. Give answers in simplest terms.

21 $\frac{1}{7}$
$- \frac{1}{8}$

22 $1\frac{1}{2}$
$+ 1\frac{3}{7}$

23 $3\frac{2}{3}$
$+ 2\frac{3}{5}$

24 $5\frac{1}{6}$
$- 4\frac{1}{7}$

25 $1\frac{1}{2}$
$+ 1\frac{2}{3}$

26 $3\frac{2}{5}$
$- 2\frac{1}{4}$

27 $8\frac{1}{7}$
$- 7\frac{1}{8}$

28 $5\frac{1}{6}$
$+ 4\frac{1}{7}$

Name _____

Solve.

29 Randy built a pool that is 30 feet × 20 feet. Then he built a wooden deck that is 6 feet wide around all four sides of the pool. What is the area of the deck? Draw a diagram to help you find the answer.

_____

30 Phillip walks $4\frac{1}{2}$ blocks to school. Joyce walks $3\frac{3}{4}$ blocks to the same school. How much farther does Phillip walk than Joyce?

_____

31

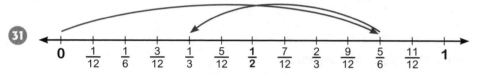

| 0 | $\frac{1}{12}$ | $\frac{1}{6}$ | $\frac{3}{12}$ | $\frac{1}{3}$ | $\frac{5}{12}$ | $\frac{1}{2}$ | $\frac{7}{12}$ | $\frac{2}{3}$ | $\frac{9}{12}$ | $\frac{5}{6}$ | $\frac{11}{12}$ | 1 |

What subtraction problem is represented on this number line?

_____

32 George recorded the height of his bean plant every week for four weeks. If the same pattern continues, how tall will his plant be in the seventh week?

_____

_____

Week	Plant Height (cm)
	2.5
1	4.25
2	6.00
3	7.75
4	9.5

33 Cammie works a total of 25 hours on Friday, Saturday, and Sunday. She works 7 hours on Saturday. She works twice as many hours on Sunday as she does on Friday. How many hours does Cammie work on Sunday?

_____

_____

34 The Romero, Hill, and Johnson families each went on a weekend outing. The Romero's parking was not the most expensive. The Hill's parking cost less than the Romero's.

Which location did each family visit?

_____

_____

_____

Weekend Outings	
Location	**Parking**
Zoo	$16.00
Swimming Pool	$15.50
Forest Preserve	$12.00

Interpreting Fractions

When you and your friends share a pizza, you may divide it and then eat, perhaps, $\frac{1}{6}$ of the pizza. You can see there is a connection between fractions and division. This lesson will introduce other ways for you to compare fractions and division.

Examples

Inez sees a drawing that represents $6 \div 3 = 2$. It also represents $6 \div 2 = 3$.

She knows from her work with improper fractions, that she can write the division problems as $\frac{6}{3} = 2$ and $\frac{6}{2} = 3$. Inez has found one way to relate fractions and division.

How can Inez make a drawing that represents $1 \div 6$?
Inez recognizes that for this problem she can draw a square.
Then she divided the square into 6 equal parts.

Inez says that if she divides the square into 6 equal parts, each part is one sixth, $\frac{1}{6}$, of the whole. Finally, she reasons that $1 \div 6 = \frac{1}{6}$. This makes the fraction an example of division.

How can Brad describe the relationship between division and fractions in one sentence?

Brad says that the 3 circles are divided into 24 parts. He writes the division expression $3 \div 24$.
Then he writes the fraction $\frac{3}{24}$. Brad said: "I can

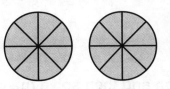

see that this fraction is related to the division of the circles into parts." He also knew that this fraction can be reduced to $\frac{1}{8}$.

Solve

Complete the fraction or division shown in the drawing.

1 $\dfrac{1}{5}$

2 $\dfrac{1}{}$

3 $3 \div \underline{}$

Sketch a drawing to show the following divisions.

4 $\dfrac{15}{3}$

5 $1 \div 12$

Name _____

Word Problems with Fractions and Mixed Numbers

Be sure to reduce any fractions to their lowest terms and rewrite improper fractions as mixed numbers.

Examples

Stringbean Farms lets visitors pick their own fruits and vegetables. Nine people spent an hour picking strawberries. They picked 24 quarts in all. If they shared the strawberries they picked equally, how many quarts of strawberries did each person take home?

Jo writes the fraction $\frac{24}{9}$ to describe how many quarts each person took home. Then she works the following division problem: $24 \div 9 = 2\frac{6}{9}$. To check her work, Jo makes a sketch:

Once Jo is sure of her answer, she writes the solution in their lowest terms.

Each person took home $2\frac{6}{9} = 2\frac{2}{3}$ quarts of strawberries.

Solve

Write a fraction and then solve the problem.

1. Four friends plan to share a 5-pound block of cheese. If they share evenly, how much cheese will each friend get? Sketch a model of the problem.
   DRAWING

   Each will receive $1\frac{1}{4}$ pounds of cheese: $\frac{5}{4} = 1\frac{1}{4}$

2. There are 21 sheep on a farm. The farmer would like to have the same number of sheep in each of his 4 fields. Is this possible? Explain why or why not.

   _____

3. Sheryl has 12 pounds of blueberries. She divides the blueberries evenly into 24 containers. What is the weight of the blueberries in each container?

   _____

4. Five people move 82 bags of chicken feed. How many bags of chicken feed will each person move if they all move the same number of bags?

   _____

Name _____

Multiplying Fractions

This lesson is an introduction to the multiplication of fractions and what the product represents.

Example

Three fifths of Lisa's drama class is in the school play. Two thirds of the actors have speaking parts they must memorize. What fraction of the drama class has a speaking part in the play?

To find $\frac{2}{3}$ of $\frac{3}{5}$, you multiply $\frac{2}{3}$ by $\frac{3}{5}$

A sketch will help.

The entire array represents the drama class. You can show $\frac{3}{5}$ of the whole to represent those in the school play. Then show $\frac{2}{3}$ of the whole to represent actors who have speaking parts. You can shade the area that you counted twice. This area represents $\frac{2}{3} \times \frac{3}{5}$.

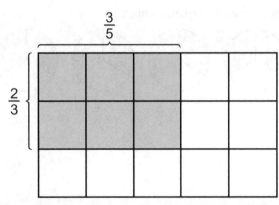

$\frac{2}{3} \times \frac{3}{5}$ is also $\dfrac{\text{the number of rectangles counted twice}}{\text{total number of rectangles}}$ = $\dfrac{2 \times 3}{3 \times 5} = \dfrac{6}{15} = \dfrac{2}{5}$

Two fifths of the students in the drama class have a speaking part in the play.

Solve

Complete the sketch. Find the product. Write your answer in their lowest terms.

1. Three fourths of the drama class are boys. Two thirds of the boys wear glasses. What fraction of the drama class represents boys wearing glasses?

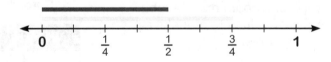

One half of the drama class are boys wearing glasses: $\frac{3}{4} \times \frac{2}{3} = \frac{6}{12} = \frac{1}{2}$

_____

2. Two thirds of the actors in the play wear costumes. One fourth of the actors' costumes need to be repaired. What fraction of the actors' costumes need to be repaired?

_____

3. Five sixths of the students in the drama class take a bus to school. Three fourths of those students take the same bus to school. What fraction of the drama class takes the same bus to school?

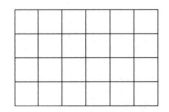

_____

Name _____

Multiplying a Whole Number by a Fraction

This lesson shows how to model the multiplication of a whole number by a fraction.

Examples

Patti makes 4 gallons of red salsa for the farmer's market. She also makes $\frac{2}{3}$ as much green salsa as red salsa. What are some models Patti can use to represent how much green salsa she should make?

First Patti draws 4 circles to represent the red salsa. Then she colors $\frac{2}{3}$ of each circle to represent the green salsa.

Next, Patti draws an array that is 4 units tall and 3 units wide. The 4 units represent the red salsa. To represent the green salsa, Patti colors $\frac{2}{3}$ of the units.

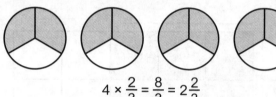

$$4 \times \frac{2}{3} = \frac{8}{3} = 2\frac{2}{3}$$

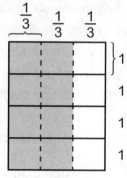

$$4 \times \frac{2}{3} = \frac{8}{3} = 2\frac{2}{3}$$

Patti also makes a model using a number line marked from 0 to 4. She marked each whole unit into thirds.

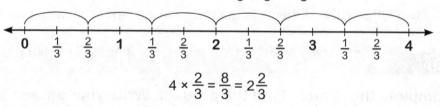

$$4 \times \frac{2}{3} = \frac{8}{3} = 2\frac{2}{3}$$

Finally, Patti decided she could make a table of values to model the problem.

Rule: Multiply by $\frac{2}{3}$

IN	OUT
0	$0 \times \frac{2}{3} = 0$
1	$1 \times \frac{2}{3} = \frac{2}{3}$
2	$2 \times \frac{2}{3} = \frac{4}{3} = 1\frac{1}{3}$
3	$3 \times \frac{2}{3} = \frac{6}{3} = 2$
4	$4 \times \frac{2}{3} = \frac{8}{3} = 2\frac{2}{3}$

Compare the products for each of these multiplications. The product is always the same!

Patti made $2\frac{2}{3}$ gallons of green salsa for the farmer's market.

$$4 \times \frac{2}{3} = \frac{8}{3} = 2\frac{2}{3}$$

Name _____

Solve

Find the product. Use any model you like.

1 $4 \times \frac{1}{8}$ _____ $\frac{4}{8} = 2$ _____

2 $6 \times \frac{1}{2}$ _____

3 $12 \times \frac{1}{2}$ _____

4 $9 \times \frac{1}{3}$ _____

5 $5 \times \frac{1}{10}$ _____

6 $3 \times \frac{1}{3}$ _____

7 $24 \times \frac{1}{3}$ _____

8 $32 \times \frac{1}{4}$ _____

9 $45 \times \frac{1}{9}$ _____

Multiply.

10 $12 \times \frac{2}{3}$ _____

11 $15 \times \frac{3}{5}$ _____

12 $8 \times \frac{3}{4}$ _____

13 $21 \times \frac{2}{7}$ _____

14 $6 \times \frac{2}{3}$ _____

15 $39 \times \frac{2}{3}$ _____

16 $24 \times \frac{3}{8}$ _____

17 $30 \times \frac{3}{5}$ _____

18 $\frac{3}{11} \times 44$ _____

Solve.

19 Maynard's Mustard prepares 14 bottles of mustard for the farmer's market. They also prepare $\frac{3}{7}$ as much ketchup as mustard. How many bottles of ketchup do they prepare?

_____

20 Lydia made calzones for the market. She started with 5 cans of tomato sauce. She used $\frac{2}{5}$ of the cans for the first batch of calzones. How many cans of tomato sauce did she use?

_____

21 From start to finish, the bakers at Bountiful Bread spend 10 hours making bread. They spend $\frac{5}{6}$ of the time waiting for the bread to rise. How much time do they wait for the bread to rise?

_____

Multiplying a Fraction by a Fraction

Examples

Ken uses $\frac{3}{4}$ of a box of screws to make a small desk. Don notices that $\frac{2}{3}$ of the screws Ken uses are longer than 2 inches. What fraction of the screws is longer than 2 inches? Don and Ken work together to draw some models to solve this problem. Ken divides a circle into fourths and colors three of the sections. Don takes Ken's circle and divides the $\frac{3}{4}$ into thirds. Then he colors two of the thirds. They decide that $\frac{3}{4} \times \frac{2}{3} = \frac{6}{12}$ or $\frac{1}{2}$.

$\frac{1}{2}$ of the screws are more than 2 inches long.

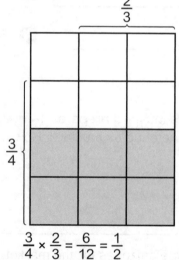

$$\frac{3}{4} \times \frac{2}{3} = \frac{6}{12} = \frac{1}{2}$$

Don draws a 4 × 3 array to describe the problem. He identifies $\frac{3}{4}$ of the array to represent the nails he used. He identifies $\frac{2}{3}$ of the array to represent the fraction of screws longer than 2 inches. The area identified twice represents the screws that were used, and were longer than 2 inches.

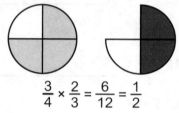

$$\frac{3}{4} \times \frac{2}{3} = \frac{6}{12} = \frac{1}{2}$$

Ken then draws a number line to model the problem. He uses a fraction strip to mark out $\frac{3}{4}$. Then he marks out $\frac{2}{3}$ of the $\frac{3}{4}$.

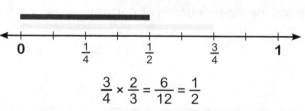

$$\frac{3}{4} \times \frac{2}{3} = \frac{6}{12} = \frac{1}{2}$$

Compare the products for each of these multiplications.

No matter which model you use, the product is always the same, $\frac{6}{12}$ or $\frac{1}{2}$.

One half of the screws are longer than 2 inches.

Name _____

Solve

Find the product. Use any model you want to help solve each problem.

1 $\frac{1}{4} \times \frac{1}{8}$ _____ $\frac{1}{32}$ _____

2 $\frac{1}{6} \times \frac{1}{2}$ _____

3 $\frac{1}{2} \times \frac{1}{2}$ _____

4 $\frac{1}{2} \times \frac{2}{3}$ _____

5 $\frac{1}{3} \times \frac{3}{4}$ _____

6 $\frac{1}{4} \times \frac{4}{5}$ _____

7 $\frac{2}{3} \times \frac{1}{3}$ _____

8 $\frac{3}{5} \times \frac{1}{5}$ _____

9 $\frac{3}{4} \times \frac{4}{5}$ _____

Multiply.

10 $\frac{5}{12} \times \frac{2}{3} =$ _____

11 $\frac{2}{15} \times \frac{3}{5} =$ _____

12 $\frac{3}{8} \times \frac{3}{8} =$ _____

13 $\frac{4}{5} \times \frac{4}{5} =$ _____

14 $\frac{5}{6} \times \frac{2}{3} =$ _____

15 $\frac{5}{12} \times \frac{3}{4} =$ _____

16 $\frac{2}{7} \times \frac{7}{12} =$ _____

17 $\frac{3}{10} \times \frac{1}{5} =$ _____

18 $\frac{3}{11} \times \frac{1}{4} =$ _____

19 $\frac{3}{10} \times \frac{1}{4} =$ _____

20 $\frac{3}{8} \times \frac{1}{4} =$ _____

21 $\frac{3}{5} \times \frac{1}{4} =$ _____

Solve. Use any model you want to help you solve.

22 Ed used $\frac{4}{5}$ of a box of floor tiles. Two thirds of the tiles he used are green. What fraction of the tiles is green?

_____

23 Marge replaces $\frac{2}{3}$ of the water pipes in her home. Three eighths of the pipe she uses is copper. What fraction of the pipe is copper?

_____

24 One half the wood trim in a dining room is oak. Five sixths of the oak trim is recycled wood. What fraction of the wood trim is recycled?

_____

25 Max used $\frac{3}{8}$ of a roll of painter's tape on the first floor of the house. About $\frac{1}{4}$ of that tape was used in the kitchen. How much of the tape was used in the kitchen?

_____

26 Kyle sketched this drawing on a piece of paper. What multiplication problem was he trying to model?

_____

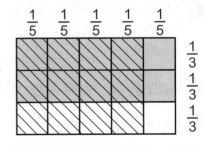

Name _____

Finding Area with Fractions

Suppose you wanted to cover a counter with hand-painted tiles. You can use what you learned about multiplying fractions to find out how many tiles you need.

Example

Maria wants to cover the top of an old counter with ceramic tiles. The top of the counter is 4 feet long and $\frac{1}{2}$ foot wide. How many square feet of tiles does Maria need to cover the counter?

Draw a grid.

Then color $\frac{1}{2}$ of each of 4 grid squares.

Finally, add up the colored grid squares: $\frac{1}{2} + \frac{1}{2} + \frac{1}{2} + \frac{1}{2} = 2$.

You can see that this is the same as multiplying $4 \times \frac{1}{2} = \frac{4}{2} = 2$

Maria needs 2 square feet of tiles to cover the counter.

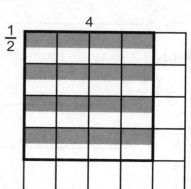

Solve

Draw a grid. Then find the area of each rectangle.

1. 4 units wide and $\frac{3}{4}$ unit tall

   area = $\frac{12}{4}$ = 3 square units

2. 6 units tall and $\frac{1}{3}$ unit wide

   area = _____

3. 4 units long and $\frac{1}{8}$ unit wide

   area = _____

4. 1 unit long and $\frac{1}{3}$ unit wide

   area = _____

Multiply to find the area of each rectangle.

5. 5 yd long × $\frac{2}{3}$ yd wide   area = _____

6. 8 m long × $\frac{2}{5}$ m wide   area = _____

Comparing Product Sizes

When you multiply, you can estimate the product by comparing the sizes of the factors.

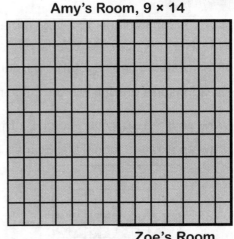

Amy's Room, 9 × 14

Zoe's Room

Example

Two sisters, Amy and Zoe, are comparing the sizes of their bedrooms. Amy claims her room is twice as large as Zoe's and that she can prove it without doing any multiplication. How do you think she does this?

Amy's bedroom is 9 feet by 14 feet. Zoe's room is 9 feet wide but only half as long as Amy's room. Draw a sketch for each room to compare the two rooms.

From the sketch you can conclude that the area of Zoe's room is $\frac{1}{2}$ the area of Amy's room. You can write an equation to describe the difference in area.

$(9 \times 14) - (9 \times \frac{14}{2}) = 126 - 63 = 63$

Solve

Describe how the products compare. Do not multiply. Explain your answer.

1. How will 4 × 12 compare to 4 × 6?

   Description: It will be _____twice_____ as great.

   Explanation: 12 is twice as great as _____.

2. How will 8 × 10 compare to 8 × 5?

   Description: It will be _____ as _____.

   Explanation: 10 is _____ as great as _____.

3. How will 30 × 24 compare to 30 × 6?

   Description: It will be _____ as _____.

   Explanation: 24 is _____ than _____.

Describe how the product sizes compare. Do not multiply.

4. How will 35 × 35 compare to 35 × 70? _____

5. How will 25 × 55 compare to 5 × 55? _____

6. How will 20 × 90 compare to 100 × 90? _____

7. How will 12 × 60 compare to 6 × 60? _____

8. How will 500 × 473 compare to 1,500 × 473? _____

Name _____

Multiplying by Fractions Greater Than or Less Than 1

You can estimate the product when you multiply a number by a fraction greater or less than 1.

Example

Ms. Johnson is planning a vegetable garden. Her garden will be 8 yards long and $\frac{2}{3}$ yard wide. Her neighbor, Ms. Evans, plans a garden 8 yards long and $\frac{3}{2}$ yards wide. How do the areas of these two vegetable gardens compare?

Ms. Johnson's garden will be smaller than Ms. Evans's garden, but how much smaller?

Draw an 8 × 3 array and color $\frac{2}{3}$ of the array to model Ms. Johnson's garden.

Draw an array to model Ms. Evans's garden.

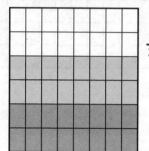   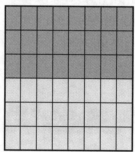

Ms. Johnson's garden:
$$8 \times \frac{2}{3} = \frac{16}{3} = 5\frac{1}{3}$$

Ms. Evans's garden:
$$8 \times \frac{3}{2} = \frac{24}{2} = 12$$

Ms. Johnson's garden is $5\frac{1}{3}$ square yards in area. Multiply 8 by a fraction less than 1 and the product is less than 8.

Ms. Evans's garden is 12 square yards in area. Multiply 8 by a fraction greater than 1 and the product is greater than 8.

Solve

Tell whether the product will be less than or greater than the whole number being multiplied. Then multiply to prove your answer.

1 $4 \times \frac{6}{5}$

Estimate: _____ more than 4 _____

Product: _____

2 $5 \times \frac{7}{6}$

Estimate: _____

Product: _____

3 $7 \times \frac{3}{3}$

Estimate: _____

Product: _____

4 $7 \times \frac{5}{3}$

Estimate: _____

Product: _____

Name _____

Tell whether the product will be less than or greater than the whole number being multiplied. Then multiply to prove your answer.

5 $4 \times \dfrac{5}{6}$

Estimate: _____

Product: _____

6 $9 \times \dfrac{7}{8}$

Estimate: _____

Product: _____

7 $11 \times \dfrac{15}{16}$

Estimate: _____

Product: _____

8 $9 \times \dfrac{3}{16}$

Estimate: _____

Product: _____

9 $9 \times \dfrac{7}{16}$

Estimate: _____

Product: _____

10 $9 \times \dfrac{15}{16}$

Estimate: _____

Product: _____

Draw arrays to model the following and then multiply.

11 Urban Steel Company is loading several old steel beams. The beams are 12 feet long and $\dfrac{4}{3}$ feet wide. Their competitor, Second Story Steel Company, is also loading several steel beams. Second Story Steel's beams are 12 feet long and $\dfrac{3}{4}$ foot wide. How do the areas of these steel beams compare? Label your arrays and show your work.

DRAWING

Name _____

Word Problems with Multiplying Fractions and Mixed Numbers

Examples

Over the weekend Fresh Produce Company sold 5 cases of mangos. On Saturday, the store sold $\frac{3}{4}$ times as many mangos as they sold on Sunday. How many mangos did the store sell on Saturday?

Randy solves the problem in three different ways. First, he draws 5 circles to represent the cases of mangos. Then he colors $\frac{3}{4}$ of each circle to represent the mangos sold.

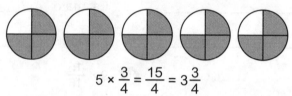

$$5 \times \frac{3}{4} = \frac{15}{4} = 3\frac{3}{4}$$

Next, Randy draws a 5 units tall and 4 units wide array. The 5 units represent the cases of mangos. Randy colors $\frac{3}{4}$ of the units to represent the mangos sold on Saturday.

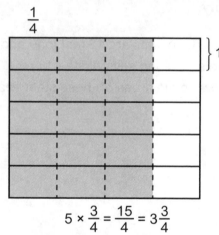

$$5 \times \frac{3}{4} = \frac{15}{4} = 3\frac{3}{4}$$

Finally, Randy made a table of values to solve the problem.

Rule: Multiply by $\frac{3}{4}$

IN	OUT
1	$1 \times \frac{3}{4} = \frac{3}{4}$
2	$2 \times \frac{3}{4} = \frac{6}{4} = 1\frac{1}{2}$
3	$3 \times \frac{3}{4} = \frac{9}{4} = 2\frac{1}{4}$
4	$4 \times \frac{3}{4} = \frac{12}{4} = 3$
5	$5 \times \frac{3}{4} = \frac{15}{4} = 3\frac{3}{4}$

$$5 \times \frac{3}{4} = \frac{15}{4} = 3\frac{3}{4}$$

Solve

Solve the problem using the method you prefer. Show your work.

1. One booth at the farmer's market started the day with $4\frac{3}{5}$ bushels of apples. By the end of the day, $\frac{5}{6}$ of the apples were sold. How many bushels of apples were sold?

Name _____

Problem Solving
You can create a table to help solve a problem.

Example

Max wants to know how many hours a week his classmates use the computer. He asks 18 classmates and gets the following responses: 2, 4, 6, 3, 5, 5, 6, 6, 2, 1, 8, 4, 5, 3, 4, 9, 4, and 3 hours. Do more students use the computer 0–2 hours or 3–5 hours a week?

Use what you know and make a table to solve the problem.
- Use intervals you think will help describe the problem: For instance, 0–2, 3–5, 6–8, 9–11.
- Make each interval the same size.
- Write tally marks to record the responses.
- Count the tallies and list the number in the *Number of Students* column.
- Compare the Numbers of Students: 10 > 4 > 3 > 1.

Hours	Tally	Number of Students
0–2	///	3
3–5	### ###	10
6–8	////	4
9–11	/	1

Max found that more students use the computer 3–5 hours a week.

Look back and check to be certain your solution makes sense.

Solve

1 DeWitt surveys his friends to see how many music videos they each downloaded in a month. DeWitt got the following responses: 0, 3, 6, 7, 9, 2, 5, 3, 8, 0, 10, 3, 5, 7, and 9

Do most of his friends download 0–2 videos, 3–5 videos, 6–8 videos, or 9–11 videos a month?

_____

Make a table to find the solution.

2 Use the data from exercise 1.

Which two intervals have the same number of responses? _____

What is that number? _____

Name _____

Write the following as division problems. Find the quotient.

1 $\frac{4}{2}$ _____

3 $\frac{8}{5}$ _____

2 $\frac{5}{3}$ _____

4 $\frac{12}{3}$ _____

Multiply whole numbers by fractions.

5 $6 \times \frac{1}{8} =$ _____

6 $4 \times \frac{1}{2} =$ _____

7 $12 \times \frac{1}{3} =$ _____

8 $9 \times \frac{2}{3} =$ _____

9 $5 \times \frac{3}{10} =$ _____

10 $4 \times \frac{3}{4} =$ _____

Find the products.

11 $\frac{1}{4} \times \frac{1}{8} =$ _____

12 $\frac{1}{6} \times \frac{1}{12} =$ _____

13 $\frac{1}{2} \times \frac{1}{2} =$ _____

14 $\frac{2}{3} \times \frac{2}{3} =$ _____

15 $\frac{2}{3} \times \frac{3}{4} =$ _____

16 $\frac{3}{4} \times \frac{4}{5} =$ _____

Tell whether the product will be less than or greater than the whole number being multiplied. Then multiply to prove your answer.

17 $7 \times \frac{7}{6}$

Estimate: _____

Product: _____

18 $7 \times \frac{6}{7}$

Estimate: _____

Product: _____

19 $9 \times \frac{4}{3}$

Estimate: _____

Product: _____

20 $9 \times \frac{13}{14}$

Estimate: _____

Product: _____

21 $11 \times \frac{1}{99}$

Estimate: _____

Product: _____

22 $11 \times \frac{98}{99}$

Estimate: _____

Product: _____

Name _____

Solve.

23 Ricky takes a survey of fifth graders for his Health and Nutrition class. He finds that in his class of 24 students, $\frac{5}{8}$ drink more than one glass of milk a day. How many students drink more than one glass of milk a day?

_____

24 Penny takes the same survey of sixth graders. She finds that in her class of 29 students, $\frac{3}{4}$ drink more than one glass of milk a day. How can Penny describe her results?

_____

25 Tyler records how many glasses of milk a group of 20 students drink in a day.

He writes the following results on a sheet of paper: 2, 4, 3, 3, 4, 2, 2, 3, 1, 4, 4, 2, 3, 2, 4, 3, 4, 3, 4, and 1. Do most of these students drink 1–2 glasses of milk or 3–4 glasses of milk? Make a table to find the solution.

_____

26 What multiplication problem is represented by these arrays?

_____

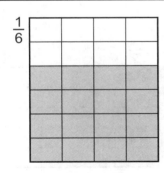
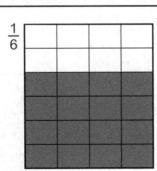
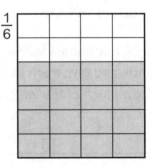

27 April plants a garden on $2\frac{1}{4}$ acres. She devotes $\frac{2}{5}$ of the entire garden to pumpkins. How many acres does April use for pumpkins? _____

28 Clint plants an apple tree and a peach tree. The apple tree is $3\frac{1}{5}$ feet tall. The peach tree is 3 times as tall as the apple tree. How tall is Clint's peach tree? _____

Interpreting Division of Fractions

A fraction with a numerator of 1 is a unit fraction. You can use models to help you divide a unit fraction by a whole number.

Example

Seina has $\frac{1}{4}$ of a bag of grapes. She wants to share them equally with herself and two friends. How much of the bag of grapes does each person get?

Step 1: This model represents the bag of grapes. The columns show $\frac{1}{4}$. Each column is divided into 3 shares. One box is one share.

$$\frac{1}{4} \quad \frac{1}{4} \quad \frac{1}{4} \quad \frac{1}{4}$$

$$\frac{1}{4} \div 3$$

Step 2: Count the boxes in the model. There are 12 boxes. Each box shows $\frac{1}{12}$ of the model or $\frac{1}{12}$ of the bag of grapes.

$$\frac{1}{4} \quad \frac{1}{4} \quad \frac{1}{4} \quad \frac{1}{4}$$

$$\frac{1}{4} \div 3 = \frac{1}{12}$$

Each person gets $\frac{1}{12}$ of the bag of grapes.

Interpret

Solve. Use models to help.

1. Joe has $\frac{1}{2}$ of a watermelon. He wants to give it to 5 friends and give each friend an equal amount. How much of the watermelon will each friend get?

   $$\frac{1}{10}$$

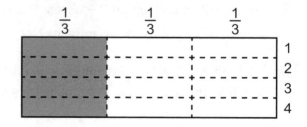

2. Kay has $\frac{1}{3}$ of a pizza. She wants to give it to 4 friends and give each friend an equal amount. How much of the pizza will each friend get?

   _____

Dividing Fractions by Whole Numbers

Example

Meg has $\frac{1}{3}$ of a bottle of juice. She wants to share it equally with herself and 4 friends. How much of the bottle of juice will each person get?

Step 1: Meg has $\frac{1}{3}$ of a bottle of juice. Draw a rectangle and divide it into three columns to show $\frac{3}{3}$. Then shade the first column to show $\frac{1}{3}$.

Step 2: Meg wants to share the juice equally with herself and 4 friends—a total of 5 people. Draw lines to show 5 rows. In the shaded column, shade one box darker.

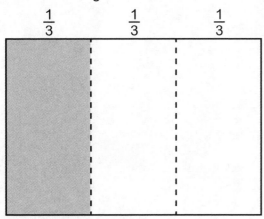

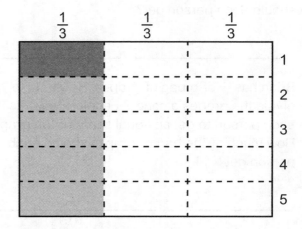

Step 3: Count the number of sections in the model. There are 15. The darkest section shows $\frac{1}{15}$ of the model or $\frac{1}{15}$ of the bottle of juice.

$$\frac{1}{3} \div 5 = \frac{1}{15}$$

Each person gets $\frac{1}{15}$ of the bottle of juice.

Step 4: Division and multiplication are inverse operations. Therefore, you can multiply to check your answer.

$$\frac{1}{15} \times \frac{5}{1} = \frac{5}{15} = \frac{1}{3}$$

The answer is correct.

Divide

Find each quotient. Use multiplication to check your answers.

1. $\frac{1}{2} \div 5 = \frac{1}{10}; \frac{1}{10} \times 5 = \frac{5}{10} = \frac{1}{2}$
   _____

2. $\frac{1}{2} \div 4 =$ _____

3. $\frac{1}{3} \div 3 =$ _____

4. $\frac{1}{4} \div 6 =$ _____

Name _____

5 $\frac{1}{5} \div 3 =$ _____

6 $\frac{1}{3} \div 5 =$ _____

Solve each problem. Draw models to help. Multiply to check your answers.

7 Yolanda has $\frac{1}{4}$ of a box of markers. She wants to divide the markers among 4 people. She wants each person to get an equal share of the markers. How much of the box of markers should each person get?

_____

8 Brian has $\frac{1}{5}$ of a bag of grapes. He wants to divide the grapes among 7 people. He wants each person to get an equal share of the grapes. How much of the bag of grapes should each person get?

_____

9 Sherri, Andy, and Grace are painting $\frac{1}{6}$ of a fence. They want to paint an equal share of the fence. How much of the fence should each person paint?

_____

10 Mona and Eli are washing $\frac{1}{2}$ of the dishes after dinner. They want to wash an equal share of the dishes. How much should each person wash?

_____

11 Jack and two friends are washing $\frac{1}{3}$ of a car. They each want to wash an equal portion of the car. How much of the car should each person wash?

_____

Interpreting Division of Whole Numbers by Fractions

You can also use models to divide a whole number by a unit fraction.

Example

Mr. Jackson has a jar that can hold 4 pints of water. If he uses a glass that can hold $\frac{1}{3}$ of a pint, how many glasses of water will he need to fill the jar?

Step 1: This model shows the problem. There are 4 columns. Each column represents 1 pint. Each column is split into 3 equal parts to show thirds.

Step 2: The quotient is the number of boxes in the model. Count the boxes. There are 12.

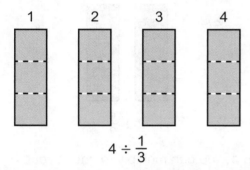

$$4 \div \frac{1}{3}$$

$$4 \div \frac{1}{3} = 12$$

Mr. Jackson needs 12 glasses of water to fill the jar.

Interpret

Solve. Use models to help.

1. Mandy has a bucket that can hold 7 gallons of liquid. If she uses a cup that can hold $\frac{1}{5}$ of a gallon, how many cups of liquid will she need to fill the bucket?

   _____ 35 _____

1	2	3	4	5	6	7
1	6	11	16	21	26	31
2	7	12	17	22	27	32
3	8	13	18	23	28	33
4	9	14	19	24	29	34
5	10	15	20	25	30	35

2. Amy has a backpack that can hold up to 5 kilograms. She also has a stack of books. Each book has a mass of $\frac{1}{3}$ of 1 kilogram. How many books will fill the backpack?

   _____

1	4	7	10	13
2	5	8	11	14
3	6	9	12	15

3. Mr. Wagner has a bowl that can hold 3 quarts. If he uses a scoop that can hold $\frac{1}{4}$ of a quart, how many scoops of liquid will he need to fill the bowl?

   _____

   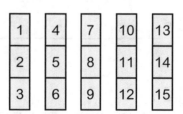

Name _____

Dividing Whole Numbers by Fractions

Example

Jamila has a mug that can hold 2 pints of tea. If she uses a spoon that can hold $\frac{1}{3}$ of a pint, how many spoonfuls will she need to fill the mug?

Step 1: Jamila's mug can hold 2 pints of tea. Draw columns to show the number of pints.

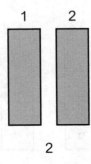

2

Step 2: Jamila uses a spoon that can hold $\frac{1}{3}$ of a pint. Draw lines to divide each column into thirds.

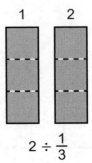

$2 \div \frac{1}{3}$

Step 3: The number of boxes you drew in the model is the quotient. Count the boxes. There are 6.

$$2 \div \frac{1}{3} = 6$$

Step 4: You can multiply to check your answer.

$$6 \times \frac{1}{3} = \frac{6}{3} = 2$$

The answer is correct.

Jamila needs 6 spoonfuls to fill the mug.

Divide

Find each quotient. Draw models to help. Use multiplication to check your answers.

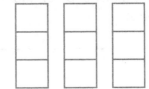

1 $3 \div \frac{1}{3} = 9; 9 \times \frac{1}{3} = \frac{9}{3} = 3$ _____

2 $5 \div \frac{1}{4} =$ _____

3 $3 \div \frac{1}{3} =$ _____

4 $7 \div \frac{1}{2} =$ _____

Solve each problem. Draw models to help. Multiply to check your answers.

5 Mr. Peters has a small bag that can hold 5 pounds of mixed nuts. If he uses a cup that holds $\frac{1}{3}$ of a pound of nuts, how many cups of nuts will he need to fill the bag?

_____

6 Harry has a bowl that can hold 4 gallons. If he uses a spoon that holds $\frac{1}{9}$ of a gallon, how many spoonfulls will it take to fill the bowl?

_____

7 Maria has a book. It will take 5 hours to read it. If she reads $\frac{1}{4}$ of an hour every day, how many days will it take for her to finish the book?

_____

8 Sam has 7 cans of paint. If he uses $\frac{1}{3}$ of a can every day, how many days will it be before he needs to buy more paint?

_____

9 A grocery store has 6 cases of grapes. If the store sells $\frac{1}{4}$ of a case of grapes every day, how many days before the store will run out of grapes?

_____

Lesson 5

Name _____

Problem Solving: Multistep Problems
Some problems will require you to complete multiple steps to find an answer.

Example

The Perez family eats $\frac{1}{2}$ pound of chicken a day. Mr. Perez went to a shop that had 56 pounds of chicken for sale. The shop had 53 pounds of chicken left after Mr. Perez bought some. How many days will it take the family to eat all the chicken Mr. Perez bought?

Step 1: Read the problem. List what you know.
The Perez family eats $\frac{1}{2}$ pound of chicken a day.
The shop had 56 pounds of chicken.
The shop had 53 pounds of chicken left after Mr. Perez bought some.

Step 2: What are you asked to find?
How many days it will take the family to eat all the chicken Mr. Perez bought.

Step 3: Write equations and solve the hidden questions.

How much chicken did the family buy?
$$56 \text{ lb} - 53 \text{ lb} = 3 \text{ lb}$$

How many days will it take the family to eat all the chicken?
$$3 \div \frac{1}{2} = 6$$

It will take 6 days for the family to eat all the chicken Mr. Perez bought.

Solve

Solve. Draw models if you need to.

1. Luke walks $\frac{1}{2}$ mile each day. Isaiah walks $\frac{1}{3}$ mile each day. How many more miles will Luke walk than Isaiah in 18 days?

   _____ 3 miles _____

2. Daisy drinks $\frac{1}{5}$ quart of juice every day. Her mom buys 15 quarts of juice. If the rest of the family drinks 9 quarts, how many days will it take Daisy to drink the rest?

   _____

3. Mr. Nelson buys 7 yards of chain that costs $18.39 per yard. Ms. Kennedy buys 8 yards of chain that costs $20.58 per yard. How much more did Ms. Kennedy spend than Mr. Nelson?

   _____

4. Mr. Sanchez has $\frac{11}{12}$ of a jug of water at the start of a soccer game. His team drinks $\frac{2}{3}$ of the water during the game. At the end of the game, Mr. Sanchez wants to divide the remaining water among a total of 3 people. He wants all 3 to get an equal share. How much of the remaining water will the 3 people get at the end of the game?

   _____

Solve.

1 Jack has $\frac{1}{2}$ of a sandwich. He wants to give it to 4 friends and give each an equal amount. How much of the sandwich will each friend get?

_____

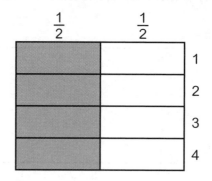

2 Tammy has $\frac{1}{4}$ of a cucumber. She wants to give it to 3 friends and give each friend an equal amount. How much of the cucumber should each friend get?

_____

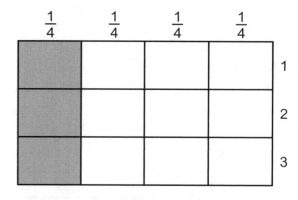

3 Maggie has a pail that can hold 4 gallons of water. If she uses a cup that can hold $\frac{1}{4}$ of a gallon, how many cups of water will she need to fill the pail?

_____

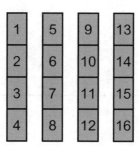

4 Amy has a box that can hold 6 kilograms of sand. If she uses a scoop that can hold $\frac{1}{5}$ of a kilogram, how many scoops of sand will she need to fill the box?

_____

1	6	11	16	21	26
2	7	12	17	22	27
3	8	13	18	23	28
4	9	14	19	24	29
5	10	15	20	25	30

5 Julio solved $\frac{1}{3} \div 5$. The quotient is $\frac{1}{15}$. Write the equation he can use to check his answer.

_____

Name _____

Find each quotient. Draw models to help. Use multiplication to check your answers.

6 $\frac{1}{4} \div 5 =$ _____

7 $\frac{1}{3} \div 6 =$ _____

8 $\frac{1}{2} \div 7 =$ _____

9 $6 \div \frac{1}{4} =$ _____

10 $4 \div \frac{1}{5} =$ _____

11 $5 \div \frac{1}{2} =$ _____

Solve.

12 The Lane family eats $\frac{1}{3}$ of a box of turkey burgers each week. Mrs. Lane goes to a store that has 72 boxes of turkey burgers for sale. The shop has 68 boxes left after Mrs. Lane buys some. How many weeks will it take the family to eat all the turkey burgers Mrs. Lane bought?

_____

Metric Units of Length

Many countries use the metric system. Common units of length are the meter, the centimeter, the millimeter, and the kilometer. Since the system is based on powers of ten, you can multiply or divide by a power of ten to convert one unit to another.

Examples

1 millimeter is about the thickness of a coin.

1 centimeter is about the width of a pen.

1 meter is about the width of a door.

1 kilometer is about 6 city blocks.

> 1 kilometer (km) = 1,000 meters (m)
> 1 meter (m) = 100 centimeters (cm)
> 1 centimeter (cm) = 10 millimeters (mm)

Mr. Brooks hiked a path that is 3.2 kilometers long. How many meters did he hike?

To change larger units to smaller units, you multiply.

$$3.2 \text{ km} \times 1,000 = 3,200 \text{ meters}$$

Ms. Michaels has some thread that is 425 millimeters long. How many centimeters of thread does she have?

To change from smaller units to larger units, you divide.

$$425 \text{ mm} \div 10 = 42.5 \text{ cm}$$

Solve

1. The Museum of Science and Industry in Chicago is about 11,600 meters from The Art Institute of Chicago. How many kilometers apart are the two buildings?

   _____ 11.6 km _____

2. Ms. Lee is 1.7 meters tall. Her daughter is 1.2 meters tall. How much taller is Ms. Lee in centimeters?

   _____

3. Doreen wants to put a model boat in a box. The model measures 252 mm tall and 459 mm wide. Can the model fit within a box that measures 28 cm tall and 32 cm wide? How do you know?

   _____

   _____

   _____

4. An adult blue whale is 25.74 meters long. How long is the blue whale in centimeters?

   _____

5. Ahmad and Eva are practicing for a bike race. Ahmad rode 3.4 km in one day. Eva rode 4.8 km the same day. How much farther did Eva ride in meters?

   _____

6. Steven has three crabapple trees in his yard. The height of one measures 610 cm. The second measures 760 cm and the third measures 910 cm. What is the sum of the heights of the trees in meters?

   _____

Name _____

Metric Units of Mass

Mass is the measure of the amount of matter in an object. Metric units used to measure mass include the milligram, the gram, the kilogram, and the metric ton. Like metric units of length, metric units of mass are based on powers of ten.

Examples

The mass of a feather is about 1 milligram.

The mass of a cherry is about 1 gram.

The mass of a pineapple is about 1 kilogram.

The mass of an adult giraffe is about 1 metric ton.

> 1 metric ton (t) = 1,000 kilograms (kg)
> 1 kilogram (kg) = 1,000 grams (g)
> 1 gram (g) = 1,000 milligrams (mg)

Ramona has an apple with a mass of 962 g. What is the mass of the apple in milligrams?

To change larger units to smaller units, you multiply.

962 g × 1,000 = 962,000 mg

A steelhead trout has a mass of 22,600 grams. What is the mass of the trout in kilograms?

To change smaller units to larger units, you divide.

22,600 g ÷ 1,000 = 22.6 kg

Solve

1. Mr. Wayne bought a watermelon with a mass of 1.89 kilograms. What is the mass of the watermelon in grams?

   _____1,890 g_____

2. Ken's dog is 4,200 grams. Sharon's dog is 7,600 grams. What is the sum of the mass of both dogs in kilograms?

   _____

3. A company has a pile of steel pipes with a total mass of 19,450 kg. The pipes need to be shipped by truck, but the company's truck can safely hold only 10 metric tons. How many trips will it take to ship all of the steel pipes? How do you know?

   _____

   _____

   _____

4. A serving of cottage cheese is considered to be 113 grams. What is the mass of a serving of cottage cheese in milligrams?

   _____

5. A science experiment calls for 130 mg of cobalt chloride and 190 mg of water. What is the difference between these two amounts in grams?

   _____

6. An elephant at a zoo has a mass of 6.8 metric tons. A rhinoceros at the same zoo has a mass of 2.2 metric tons, and a hippopotamus has a mass of 2.9 metric tons. What is the sum of the mass of these three animals in kilograms?

   _____

Name _____

Metric Units of Capacity

Capacity is the volume of a container measured in liquid units. Metric units used to measure capacity include the milliliter, the deciliter, and the liter. Metric units of capacity are based on powers of ten, like all other metric units.

Examples

A milliliter is about 20 drops of liquid from an eyedropper.

The capacity of a small cup is about 1 deciliter.

A mid-sized bottle has a capacity of about 1 liter.

> 1 liter (L) = 10 deciliters (dL)
> 1 liter (L) = 1,000 milliliters (mL)
> 1 deciliter (dL) = 100 milliliters (mL)

Alan made a bowl in pottery class. It has a capacity of 534 milliliters. What is the capacity of the bowl in deciliters?

To change smaller units to larger units, you divide.

534 mL ÷ 100 = 5.34 dL

22.8 liters of hot water are in a bathtub. How many deciliters of water are in the bathtub?

To change larger units to smaller units, you multiply.

22.8 L × 10 = 228 dL

Solve

1 Amy's aquarium holds 70 liters of water. How many milliliters of water does the aquarium hold?

_____ 70,000 mL _____

2 A red pitcher holds 20.4 deciliters of juice. A blue pitcher holds 14.7 deciliters of juice. How many more liters of juice are in the red pitcher?

_____

3 An artist needs 30 L of paint for an outdoor wall mural. He has 82 dL of red paint, 67 dL of blue paint, 32 dL of green paint, and 70 dL of yellow paint. Does he have enough paint for the mural? How do you know?

_____

_____

_____

4 Geoff drank about 220 milliliters of water. How many deciliters of water did he drink?

_____

5 Sheila is filling a 2.2-L bottle with liquid from a full 1.1-dL glass. How many glasses of liquid will it take to fill the bottle?

_____

6 Mechanics at a garage use 87 liters of oil on Monday. They use 66 liters on Tuesday, and 103 liters on Wednesday. How many milliliters of oil do they use in all?

_____

Name _____

Units of Time

There are many units of time. Some units of time are the second, the minute, the hour, the day, the week, the month, and the year. You can convert different units of time to compare them.

Examples

1 minute = 60 seconds	1 hour = 60 minutes
1 day = 24 hours	1 week = 7 days
1 month = about 4 weeks	1 year = 12 months
1 year = 52 weeks	1 year = 365 days

Elaine went on a vacation with her family for 14 days. How many weeks did Elaine spend on vacation?

To change smaller units to larger units, you divide.

14 days ÷ 7 = 2 weeks

Lorne spent 17 minutes cleaning his room. How many seconds did it take for Lorne to clean his room?

To change larger units to smaller units, you multiply.

17 minutes × 60 = 1,020 seconds

Solve

1 Jin spent $1\frac{1}{2}$ hours doing her homework. How many minutes did she spend doing homework?

_____ 90 minutes _____

2 Marsha jogged for $36\frac{3}{4}$ minutes. Gary jogged for $28\frac{3}{5}$ minutes. How much longer did Marsha jog in seconds?

_____

3 Samir spends 9 minutes a day brushing his teeth. If Samir does this every day, how many hours does he spend brushing his teeth in a year?

_____

4 A red maple tree lives for an average of 70 years. What is the lifespan of a red maple tree in weeks?

_____

5 Ms. Singh spends 72 months paying a car loan. Mr. Pedersen spends 48 months paying a car loan. How much longer does Ms. Singh pay for her loan in years?

_____

6 A box turtle named Houdini has lived at a zoo for 63 years. Another box turtle, named Kitt, has lived at the same zoo for 39 years. How many days in all have both turtles lived at the zoo? Use 365 days per year in your calculation.

_____

Name _____

Using Celsius Temperatures

Metric units of temperature are degrees Celsius (°C). A hot summer day could be 32°C. A cold winter day could be –3°C. Water freezes at 0°C. You can solve problems that involve changes in temperature using thermometers and addition or subtraction.

Example

At 7:00 A.M., Laura read a thermometer. It showed 8°C. When she read the thermometer again at 1:00 P.M., it showed 22°C. How many degrees Celsius did the temperature increase?

You can add or subtract to find changes in temperature. In this case, you can subtract to find the change in temperature between 7:00 A.M. and 1:00 P.M.

$$22°C - 8°C = 14°C$$

The temperature increased by 14°C between 7:00 A.M. and 1:00 P.M.

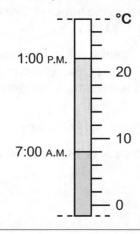

Solve

1. The high temperature for a day in July was 35°C. The low temperature that same day was 23°C less. What was the low temperature on that day?

   _____ 12°C _____

2. What temperature does the thermometer show?

   _____

   If the temperature increased by 14°C, what would the thermometer show?

   _____

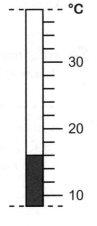

3. Kevin and Sharon live in different cities. One day, the temperature in Kevin's city was 24°C. The temperature in Sharon's city that same day was 9°C higher. What was the temperature in Sharon's city?

   _____

4. What temperature does the thermometer show?

   _____

   If the temperature decreased 25°C, what would the thermometer show?

   _____

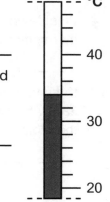

Name _____

Using Fahrenheit Temperatures

Customary units of temperature are degrees Fahrenheit (°F). A hot summer day could be 93°F. A cold winter day could be 12°F. Water freezes at 32°F. You can solve problems that involve changes in temperature using thermometers and addition or subtraction.

Example

At 7:30 A.M., Keith read a thermometer. It showed 76°F. When he read the thermometer again at 1:30 P.M., the temperature was 18°F higher. What was the temperature in degrees Fahrenheit at 1:30 P.M.?

You can add or subtract to find changes in temperature. In this case, you can add to find the change in temperature between 7:30 A.M. and 1:30 P.M.

$$76°F + 18°F = 94°F$$

The temperature was 94°F at 1:30 P.M.

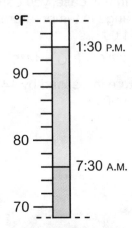

Solve

1. The high temperature for a day in November was 58°F. The low temperature that same day was 15°F less. What was the low temperature on that day?

   43°F

2. What temperature does the thermometer show?

   _____

   If the temperature increased 39°F, what would the thermometer show?

   _____

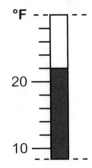

3. Jessica and Olivia live in different cities. One day, the temperature in Jessica's city was 76°F. That same day, the temperature in Olivia's city was 27°F higher. What was the temperature in Olivia's city?

   _____

4. What temperature does the thermometer show?

   _____

   If the temperature decreased 19°F, what would the thermometer show?

   _____

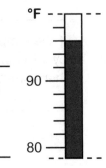

Customary Units of Length

The United States usually uses the customary system of measurement. Common units of length are the inch, the foot, the yard, and the mile. You can multiply or divide to convert one unit to another. The numbers you use when you divide or multiply depend on the units you are using.

Examples

A medium-sized paperclip is about 1 inch long.	Many textbooks are about 1 foot tall.	A baseball bat is about 1 yard in length.	Most people can walk a mile in about 20 minutes.

> 1 foot (ft) = 12 inches (in)
> 1 yard (yd) = 3 feet (ft)
> 1 mile (mi) = 1,760 yards (yd)
> 1 mile (mi) = 5,280 feet (ft)

Bob has 22 yards of kite string. How many feet of kite string does Bob have?

To change larger units to smaller units, you multiply.

$$22 \text{ yd} \times 3 = 66 \text{ ft}$$

Newhaven Street is 8,834 yards long. What is the length of Newhaven Street in miles?

To change smaller units to larger units, you divide.

$$8,834 \text{ yd} \div 1,760 = 5 \text{ R}34 \text{ mi}$$

Solve

1 One young maple tree is 7 feet 2 inches tall. A young oak tree is 108 inches tall. Which tree is taller?

_____the oak tree_____

2 There are two towers in a major city. The first tower is 1,483 feet tall. The second tower is 1,891 feet tall. How much taller is the second tower in yards?

_____

3 Joe wants to store a bicycle in a closet for the winter. The closet is 7 feet long and 4 feet wide. His bicycle is 68 inches long and 22 inches wide. Can he fit the bicycle into his closet? How do you know?

_____

_____

4 A hiking trail in a state park is 5.2 miles long. How long is the trail in feet?

_____

5 Mr. Edwards is 6 feet 3 inches tall. Mrs. Edwards is 5 feet 7 inches tall. What is the difference of their heights in inches?

_____

6 Ramona, Faith, and Anna go camping. The distance from Ramona's tent to her car is 699 feet. The distance from Faith's tent to her car is 768 feet, and the distance from Anna's tent to her car is 724 feet. What is the sum of these distances in yards?

_____

Name _____

Customary Units of Weight

Weight is the measure of how light or heavy something is. Common units of weight are the ounce, the pound, and the ton. You can multiply or divide to convert one unit to another. The numbers you use when you divide or multiply depend on the units you are using.

Examples

A slice of bread weighs about 1 ounce.

A small melon weighs about 1 pound.

A small car weighs about 1 ton.

1 pound (lb) = 16 ounces (oz)
1 ton (T) = 2,000 pounds (lb)

Ms. Arnold buys 4 pounds of potato salad for a party. How many ounces of potato salad does she buy?

To change larger units to smaller units, you multiply.

$$4 \text{ lb} \times 16 = 64 \text{ oz}$$

An elephant in a safari park weighs 14,000 pounds. What is the weight of the elephant in tons?

To change smaller units to larger units, you divide.

$$14,000 \text{ lb} \div 2,000 = 7 \text{ T}$$

Solve

1 Dan's beagle weighs 24 pounds. What is the weight of his beagle in ounces?

_____ 384 oz _____

2 A school bus weighs 17,400 pounds. A truck weighs $7\frac{1}{2}$ tons. How much heavier is the bus in pounds?

_____

3 Mona has a large straw basket that can hold up to 25 lbs without breaking. She wants to put 220 oz of apples and 160 oz of bananas into the basket. Can the basket hold the apples and bananas without breaking? How do you know?

_____

_____

_____

4 A whale at an aquarium weighs 8 tons. What is the weight of the whale in pounds?

_____

5 Ms. Hoy finds two boulders at an archaeological site. One boulder weighs 3.2 T and the other weighs 1.8 T. What is the sum of the weights of the boulders in pounds?

_____

6 Three people buy fish at a store. Mr. Jones buys 1.3 pounds, Ms. Smith buys 1.5 pounds, and Mr. Lee buys 2 pounds. How many ounces of fish do they buy in all?

_____

Name _____

Customary Units of Capacity

Capacity is the volume of a container measured in liquid units. Customary units of capacity are the cup, the pint, the quart, and the gallon. You can multiply or divide to convert one unit to another. The numbers you use when you divide or multiply depend on the units you are using.

Examples

A coffee cup holds about 1 cup.

A small carton holds about 1 pint.

Many water bottles hold about 1 quart.

Many people buy milk in 1-gallon bottles.

> 1 pint (pt) = 2 cups (c)
> 1 quart (qt) = 2 pints (pt)
> 1 gallon (gal) = 4 quarts (qt)

Mr. Nowak buys 5 quarts of motor oil. How many pints of motor oil does he buy?

To change larger units to smaller units, you multiply.

$$5 \text{ qt} \times 2 = 10 \text{ pt}$$

Lori filled a 16-quart jug with water. How many gallons of water are in the jug?

To change smaller units to larger units, you divide.

$$16 \text{ qt} \div 4 = 4 \text{ gal}$$

Solve

1 A sink has 2 gallons of water in it. How many quarts of water are in the sink?

_____8 qt_____

2 Cindy drinks 2.3 pints of water after school. Her sister drinks 1.8 pints. How many more cups of water does Cindy drink?

_____

3 Mr. Rose's car has a 12-gallon gas tank. His wife drives a car with a 10-gallon gas tank, and his daughter also drives a car with a 10-gallon gas tank. If all three tanks are filled, how many quarts of gas would they have in all? How do you know?

_____
_____
_____

4 A pitcher holds 5 quarts of lemonade. How many pints of lemonade does the pitcher hold?

_____

5 A slow-cooker pot has a capacity of 64 cups. What is the capacity of the pot in pints?

_____

What is the capacity of the pot in quarts?

_____

6 Maria's bathtub can hold 160 quarts of water. Barbara's bathtub can hold 220 quarts of water. How much more water can Barbara's bathtub hold in gallons?

_____

Name _____

Problem Solving: Extra or Missing Information

Some problems have extra information. You do not need the extra information to solve the problem. Other problems are missing information. If information is missing, then you cannot solve the problem.

Example

John planted some sunflower seeds in March. In June, he measured the height of one sunflower plant. It was 97 cm tall. In August, he measured it again. It was 3 times as tall. What is the height of the sunflower plant in August?

Step 1: Draw a diagram to show what you know.

Height of Sunflower in August?

3 times as tall | 97 cm | 97 cm | 97 cm

Height in June | 97 cm

Step 2: Be sure you understand the problem. Do you need any additional information to solve it?

No. All the information you need is given.

Is there extra information not needed to solve it?

Yes. It does not matter when John plants the sunflower seeds.

Since 97 cm × 3 = 291, the height of the sunflower was 291 cm in August.

Solve

Solve the problem if you have enough information. Tell any information that is not needed or that is missing.

1 Chloe walks about 3.8 kilometers every day. Katie walks about half as much. How many kilometers does Chloe walk every 40 days?

152 km; You do not need to know that Katie walks about half as much to solve the problem.

2 Gabe brought 1.3 liters of water, 2.19 liters of juice, and 1.8 liters of lemonade to a picnic. 19 people came to the picnic. At the end of the picnic, Gabe had 1.47 liters of drinks left over. How much did everyone drink at the picnic?

_____

_____

3 A male dolphin at a water park eats about 11.83 kg of food in a day. A female dolphin eats about 10.76 kg of food in a day. $\frac{2}{3}$ of the food they eat is fish. How much food will both dolphins eat in a week?

_____

_____

4 A science experiment calls for 13 mL of red food coloring, 2 mL of blue food coloring, and 1 mL of yellow food coloring. How many milliliters of food coloring are needed so that every student in the class can perform the experiment?

_____

_____

Problem Solving: Working Backwards

You can solve some problems by "working backwards." Writing an equation and inverse operations can help you do this.

Example

Jenna has a large roll of ribbon. She gave $14\frac{1}{4}$ ft to her brother. She also gave $9\frac{3}{4}$ ft to her father. She has 51 feet of ribbon left. How many feet of ribbon did Jenna start with?

Step 1: Write an equation to show the problem. Use variables to show unknown numbers.

$$y - 14\frac{1}{4} - 9\frac{3}{4} = 51$$

Step 2: Now you know the end number and how it was calculated. You can work backward using inverse operations to find the beginning number.

$$51 + 9\frac{3}{4} + 14\frac{1}{4} = y$$
$$y = 75$$

Jenna had 75 feet of ribbon to start with.

Solve

1 Regina and Alicia had $4\frac{1}{2}$ cups of water at the end of a day of driving. They drank $8\frac{1}{4}$ cups before lunch and $11\frac{1}{4}$ cups in the afternoon. How much water did they start with?

24 c or $1\frac{1}{2}$ gal
_____

2 A small shop has 5.9 pounds of lunchmeat at the end of Friday. The shop sold 4.9 pounds on Friday morning and 9.35 pounds on Friday afternoon. How many pounds of lunchmeat did the store start with?

_____

3 Use the digits 7, 3, 6, 5, 1, and 9 to write three numbers less than 600,000 but greater than 500,000. Use each digit only once for each number.

_____

4 Ms. Torres is on a road trip. She drives a total of 2,731.82 miles in three weeks. She drives 791.38 miles the second week. She drives 1,086.14 miles the third week. How many miles did she drive the first week?

_____

5 Mr. Kerry has 4 boxes of cocoa mix. Each box has 20 bags. He uses two cups of water and one bag of cocoa mix to make a mug of cocoa. How many cups of water will Mr. Kerry use if he uses all the bags of cocoa mix?

_____

6 Ling is $56\frac{3}{4}$ in. tall. Lauren is $49\frac{1}{3}$ in. tall. How much taller is Ling than Lauren?

_____

Name _____

Solve.

1 The high temperature for a day in August was 37°C. The low temperature that day was 19°C less. What was the low temperature on that day?

_____

2 What temperature does the thermometer show?

_____

If the temperature increased 21°C, what would the thermometer show?

_____

3 Joel and Terry live in separate towns. One day, the temperature in Joel's town was 96°F. The temperature in Terry's town was 65°F lower. What was the temperature in Terry's town?

_____

4 What temperature does the thermometer show?

_____

If the temperature decreased 13°F, what would the thermometer show?

_____

5 Anna wants to put a large bowl in a box. The bowl measures 31 cm wide and 15 cm tall. Can the bowl fit within a box that measures 340 mm wide and 200 mm tall? How do you know?

_____

_____

6 John F. Kennedy International Airport is about 27.82 km from the Empire State Building in New York City. How many meters is the airport from the Empire State Building?

_____

_____

7 Arlen bought 1.34 kilograms of apples. What is the mass of the apples in grams?

_____

_____

8 A standard serving of green grapes has a mass of 92 g. A standard serving of cheddar cheese has a mass of 28 g. What is the difference in the serving sizes in milligrams?

_____

_____

Solve.

9 A blue bottle holds $13\frac{3}{4}$ deciliters of juice. A green bottle holds $12\frac{1}{2}$ deciliters of juice. How many more milliliters of juice are in the blue bottle?

_____

10 Mr. Lee rode a bike for 53.8 minutes. Ms. Lee rode a bike for 62.1 minutes. How much longer did Ms. Lee ride a bike in seconds?

_____

11 There are two statues in a city. The first statue is $21\frac{1}{2}$ feet tall. The second statue is 35 feet tall. How much taller is the second statue in inches?

_____

12 Eloisa and her mom visit relatives. They drive 135 miles on Saturday. They drive 149 miles on Sunday. What is the total distance they traveled in yards?

_____

13 A small fish tank has 12 gallons of water in it. How many quarts are in the fish tank?

_____

14 A small pot holds 2 quarts of soup. A large pot holds 5 quarts of soup. How many more pints of soup does the large pot hold?

_____

Solve if you have enough information. Tell any information that is not needed or that is missing.

15 A small store has 32 gallons of milk to sell. The store sells 13 gallons on Monday and more on Tuesday. How many gallons of milk did the store have left?

_____

_____

16 From home, Ken ran 15 blocks. He stopped at a park for 30 minutes. Then he ran $22\frac{3}{4}$ blocks. Later he returned home following the same route. What is the total distance Ken ran?

_____

_____

Name _____

Lines and Segments

A line is a straight path through two points. It goes on in both directions without ending.

You write: $\overleftrightarrow{AB}$

Parallel lines never cross. They are always the same distance apart.

A ray has one endpoint. It goes on forever in one direction.

You write: $\overrightarrow{AB}$

Intersecting lines cross each other at a single point.

A line segment is a part of a line. It has two endpoints.

You write: $\overline{CD}$

Perpendicular lines form square corners, or right angles, where they cross.

Identify

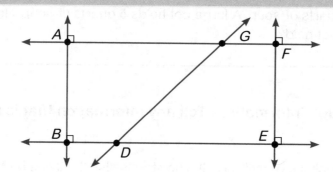

1 Identify two points.

  Point A, Point G

2 Identify two line segments.

  _____

3 Identify two parallel lines.

  _____

4 Identify two perpendicular lines.

  _____

5 Identify two intersecting lines that are not perpendicular.

  _____

Identifying and Measuring Angles

An angle is made up of two rays that meet at an end point.
The point where the two rays meet is called the vertex.

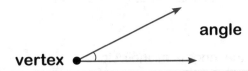

vertex   angle

Examples

There are four types of angles.

| A right angle has a measure of exactly 90°. | An acute angle has a measure between 0° and 90°. | An obtuse angle has a measure between 90° and 180°. | A straight angle has a measure of exactly 180°. |

right angle   acute angle   obtuse angle   straight angle

You use a protractor to measure angles. Angles are measured in degrees.

Step 1: Place the protractor's center on the angle's vertex.
Step 2: Align the 0 mark of one of the protractor's sides with one ray of the angle to be measured.
Step 3: Find where the other ray passes. Read the measure.
∠DFG is 30 degrees or 30°.

Identify and Measure

Classify each angle as straight, right, obtuse, or acute. Measure each angle with a protractor. You may have to extend the rays of an angle to measure it. Write each measure.

1. _____ acute; 84° _____

2. 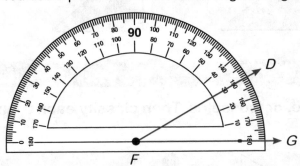 _____

3. _____

4. 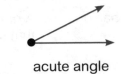 _____

5. 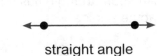 _____

6.  _____

Name _____

Classifying Triangles

Triangles have three sides and three angles. They can be classified by the lengths of their sides or by the measure of their angles.

Examples

You can classify triangles by the lengths of their sides.

An equilateral triangle has three sides that are the same, or equal length.

An isosceles triangle has two sides that are the same length.

A scalene triangle has no sides that are the same length.

An acute triangle has three acute angles.

An obtuse triangle has one obtuse angle.

A right triangle has one angle that measures 90°.

Classify

Classify each triangle as equilateral, scalene, or isosceles. Then classify each triangle as acute, obtuse, or right.

1
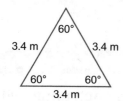

equilateral triangle, acute triangle

2
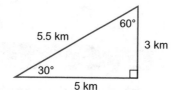
_____

3
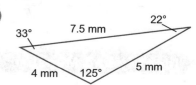
_____

4
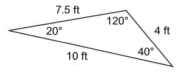
_____

5
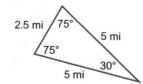
_____

6
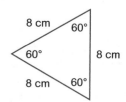
_____

Classifying Quadrilaterals

Quadrilaterals have four sides and four angles. They can be classified by their pairs of sides or by the measure of their angles.

Examples

A parallelogram has four sides. Opposite sides are parallel and equal in length.

A trapezoid has four sides. Only one pair of sides is parallel.

A rhombus is a type of parallelogram. All four sides are the same length.

A rectangle is a type of parallelogram. Each corner forms a right angle, and its opposite sides are the same length.

A square is a rectangle with four sides all the same length. It is also a rhombus.

Classify

Classify each quadrilateral.

1

_____trapezoid_____

2

6 ft · 90° · 6 ft · 90° · 90° · 6 ft · 90° · 6 ft

_____

3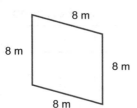

8 m · 8 m · 8 m · 8 m

_____

4

_____

5 Hugo says that all squares have four right angles because all squares are rhombi and all rhombi have four right angles. Do you agree? Why or why not?

_____

Attributes of Polygons

A plane is an endless flat surface. A polygon is a closed plane figure. Its sides are line segments.

Examples

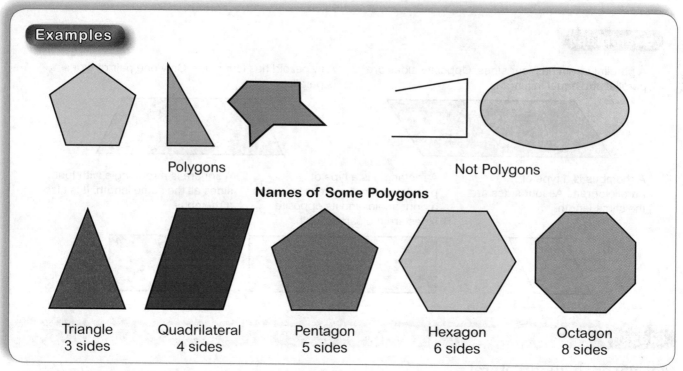

Polygons   Not Polygons

Names of Some Polygons

| Triangle 3 sides | Quadrilateral 4 sides | Pentagon 5 sides | Hexagon 6 sides | Octagon 8 sides |

Categorize and Identify

1   Look at the row of figures. Which figures are polygons? Which are not?

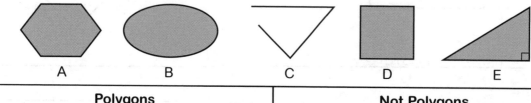

A   B   C   D   E

Polygons	Not Polygons
A, D, and E are polygons	B and C are not polygons

Name each polygon.

2    3   4

_____   _____   _____

5   Is a circle a polygon? Why or why not?

_____

Classifying Polygons

A regular polygon has sides of equal length and angles of equal measure. An irregular polygon has sides of unequal length and angles of unequal measure.

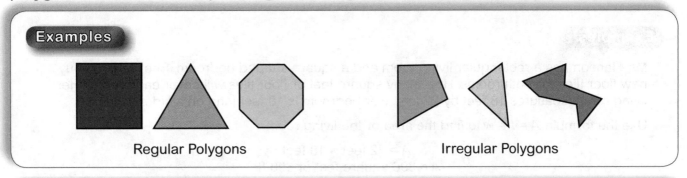

Examples

Regular Polygons   Irregular Polygons

Identify and Classify

Identify each plane shape as regular or irregular.

_____irregular_____

❷

_____

❸

_____

The flowchart below classifies triangles and quadrilaterals. Fill in the blanks with the correct term.

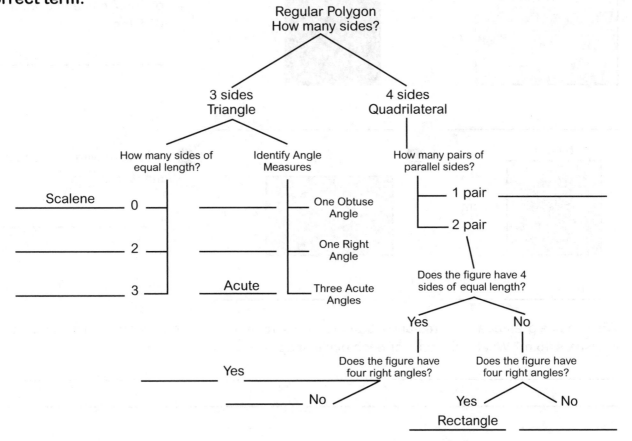

Regular Polygon
How many sides?

3 sides
Triangle

4 sides
Quadrilateral

How many sides of equal length?

Identify Angle Measures

How many pairs of parallel sides?

Scalene _____ 0

_____ 2

_____ 3

_____ One Obtuse Angle

_____ One Right Angle

Acute _____ Three Acute Angles

1 pair _____

2 pair

Does the figure have 4 sides of equal length?

Yes   No

_____ Yes _____

_____ No

Does the figure have four right angles?

Does the figure have four right angles?

Yes   No

Rectangle _____

Name _____

Finding the Area of a Rectangle

You can use the formula $A = l \times w$ to find the area of a rectangle. You can also use the formula $A = s \times s$ to find the area of a square.

Examples

Ms. Hanson has a rectangular living room and a square-shaped bedroom. She wants to buy new floor tiles for both rooms. How many square feet of floor tiles will cover each room? Her living room measures 18 feet by 22 feet. Her bedroom is 18 feet long on a side.

Use the formula $A = l \times w$ to find the area of the living room.

$$A = 22 \text{ feet} \times 18 \text{ feet}$$
$$A = 396 \text{ square feet or } 396 \text{ ft}^2$$

The length and the width of the bedroom are the same since it is a square shape. Use the formula $A = s \times s$ to find the area.

$$A = 18 \text{ feet} \times 18 \text{ feet}$$
$$A = 324 \text{ square feet or } 324 \text{ ft}^2$$

Multiply

Find the area of each shape.

1

9 m

4 m

_____ 36 m² _____

2

74.2 km

74.2 km

_____

3 A rectangle with a length of 105.7 mm and a width 66.8 mm

_____

4

64 cm

64 cm

_____

5

$81\frac{3}{4}$ in.

$47\frac{1}{4}$ in.

_____

6 A square with a side that measures $\frac{4}{5}$ mi

_____

7 Which has a greater area: a rectangle that measures 15 m by 17 m or a square with a side that measures 16 m? What is the area of each plane shape?

_____

_____

Name _____

Counting Unit Cubes and Volume

You can use cubic units to fill solid figures. The number of cubic units needed to fill a solid figure is called volume.

Examples

A cubic unit measures 1 unit high, 1 unit wide, and 1 unit long. It has a volume of 1 cubic unit or 1 unit^3.

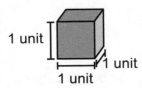

1 unit

1 unit

1 unit

You can count cubes to find volume of a solid object if the cubes are visible. Start with the base.

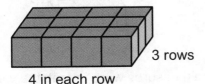

3 rows

4 in each row

4 + 4 + 4 = 12

You can often use cubic units to measure volume, since they can be stacked without gaps or overlaps.

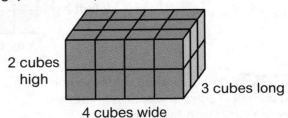

2 cubes high

3 cubes long

4 cubes wide

Then add the number of cubes in each layer.

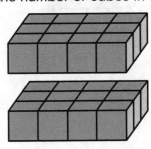

12 + 12 = 24 cubes

Count

Count the cubes in each object. Write the total.

1 _____8_____

2 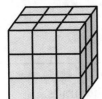 _____

3 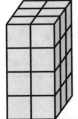 _____

4 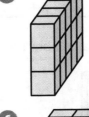 _____

5 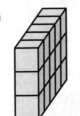 _____

6  _____

Name _____

Measuring Volume with Unit Cubes

Example

Ms. Vega has a large box that measures 3 feet long, 4 feet wide, and 2 feet high. What is the volume of the box?

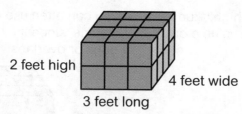

2 feet high

4 feet wide

3 feet long

Remember! You can count cubes to find volume of a solid object if the units are visible. Start with the base.

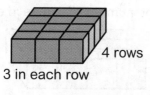

4 rows

3 in each row

3 + 3 + 3 + 3 = 12

Then add the number of cubes in each layer.

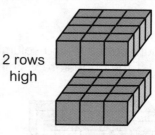

2 rows high

12 + 12 = 24 cubes

Since the measures were in feet, the volume of the box is 24 cubic feet or 24 ft³.

Count

Count the cubes in each solid object. Write the volume.

❶

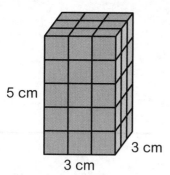

5 cm

3 cm

3 cm

_____45 cm³_____

❷

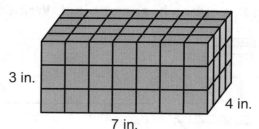

3 in.

7 in.

4 in.

_____

❸ Look at the cubes below. Write a number sentence to find the volume. Use specific units.

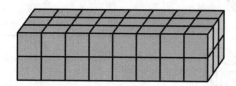

_____

Multiplying Using Unit Cubes

Example

What is the volume of a rectangular box that measures 7 in. wide, 7 in. long, and 6 in. high?

Remember! $A = l \times w$

Counting the cubes in a square or rectangular base is similar to finding the area of a square or rectangular base.

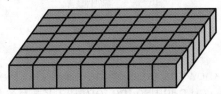

$$7 + 7 + 7 + 7 + 7 + 7 + 7 = 49$$
7 rows (width) × 7 cubes in each row (length)
= 49 cubes

Counting the cubes in each layer is similar to multiplying the area of the base by the height of the figure.

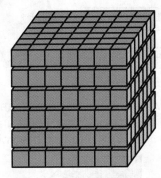

$$49 + 49 + 49 + 49 + 49 + 49 = 294 \text{ cubes}$$

6 layers (height) × 49 cubes in each layer
(base area) = 294 cubes

You can find the volume of a square or rectangular object by multiplying the area of the base by the height of the object, or $(l \times w) \times h$.

The box is $(7 \times 7) \times 6 = 294$ cubic inches.

Multiply

Find the volume of each object. Write the equation you used.

 1

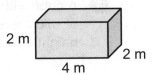
2 m
4 m
2 m

_____$(4 \times 2) \times 2 = 16 \text{ m}^3$_____

 2

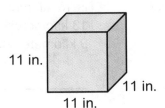
11 in.
11 in.
11 in.

_____

3

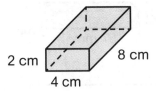
2 cm
4 cm
8 cm

_____

4

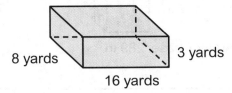
8 yards
16 yards
3 yards

_____

Name _____

Finding Volume Using Formulas

A rectangular prism is a solid figure with six faces that are rectangles or squares. A cube is a solid figure with six faces that are just squares.

Examples

Some Rectangular Prisms

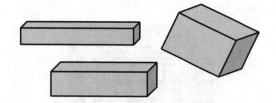

If you know the length, the width, and the height of a prism or a cube, you can use the formula $V = (l \times w) \times h$ to find the volume.

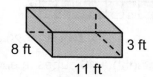

8 ft   3 ft
11 ft

$$V = (l \times w) \times h$$
$$V = (8 \times 11) \times 3$$
$$V = 264 \text{ ft}^3$$

Cube

If you know the area of the base and the height, you can use the formula $V = B \times h$ to find the volume.

6 m   Area of Base: 36 m²

$$V = B \times h$$
$$V = 36 \times 6$$
$$V = 216 \text{ m}^s$$

Calculate

Find the volume of each object. Use $V = B \times h$ or $V = (l \times w) \times h$.

1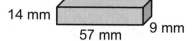

14 mm
57 mm   9 mm

_____7,182 mm³_____

2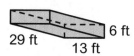

29 ft   6 ft
13 ft

_____

3 A cube with a base area of 676 square miles and a height of 26 miles

_____

4

17 in.

Area of base: 289 in.²

_____

5 A rectangular prism that is 7 yards high and a base area of 84 square yards

_____

6 A rectangular prism that is 15 kilometers high, 13 kilometers wide, and 9 kilometers long

_____

Finding the Volume of Irregular Solids

Many irregularly shaped solids can be split into smaller parts.

Paula finds an oddly shaped box. She wants to know the volume of the box. How can she find the volume?

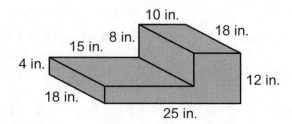

Step 1: Separate the solid into two rectangular prisms. Find the length, width, and height of each.

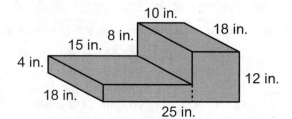

Step 2: Find the volume of each solid.

$$V = (l \times w) \times h \qquad V = (l \times w) \times h$$
$$V = (18 \times 15) \times 4 \qquad V = (18 \times 10) \times 12$$
$$V = 1{,}080 \text{ in.}^3 \qquad V = 2{,}160 \text{ in.}^3$$

Step 3: Add the volumes. $1{,}080 + 2{,}160 = 3{,}240 \text{ in.}^3$
The volume of the oddly shaped box is $3{,}240 \text{ in.}^3$.

Find the volume of each object.

1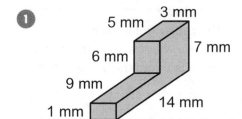

147 mm³

3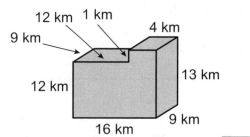

_____

2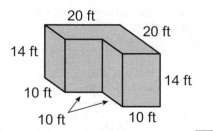

_____

4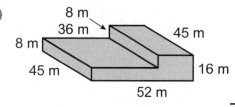

_____

Name _____

Problem Solving: Using Formulas

Example

A warehouse receives a shipment of boxes. Each box is cube-shaped and 3 feet high. If the boxes are stacked 4 high, 5 deep, and 4 wide, what will be the total volume of the shipment of boxes?

Step 1: What do you know?

Each box is shaped like a cube.
Each box is 3 feet high.
The boxes will be stacked 4 high, 5 deep, and 4 wide.

What are you asked to find?

The total volume of the shipment

Step 2: Think of how to find a solution.

Find the height of 4 boxes, the length of 5 boxes, and the width of 4 boxes. Then use the formula for volume to find the total volume of the shipment.

Step 3: Solve. Use the formula $V = (l \times w) \times h$.

Height of the Stack = 3 feet × 4 high = 12 feet
Length of the Stack = 3 feet × 5 deep = 15 feet
Width of the Stack = 3 feet × 4 wide = 12 feet
Volume of the Stack = (15 × 12) × 12 = 2,160 ft^3

The total volume of the shipment will be 2,160 ft^3.

Solve

Solve. Use the formulas $V = (l \times w) \times h$, $V = B \times h$, or $A = l \times w$ if necessary.

1 A large fish pond is 5 meters long, 4 meters wide, and 12 centimeters deep. What is the volume of the pond in centimeters?

_____ 2,400,000 cm^3 _____

2 A rectangular pool has a base area of 600 ft^2. The total volume of the pool is 4,800 ft^3. How deep is the pool?

_____

3 A store sells dragon figures in 10-inch cube boxes. How many boxes of figures can fit on a shelf that is 30 inches long, 20 inches high, and 30 inches wide?

_____

4 Dylan covers the rectangular floor of his room with 24 square meters of carpet. The length of the room is 6 meters. What is the width of the room?

_____

Use the table for exercise 5.

5 How much greater in volume is the backpack than the tote?

_____

Jack's Bestbuy Products

Model	Dimensions (in.)
Tote	16 × 15 × 12
Backpack	20 × 15 × 10
Suitcase	24 × 24 × 12

Use the diagram for exercises 1–3.

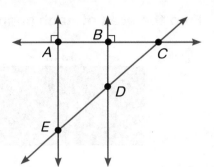

1 Identify two line segments.

_____

2 Identify two parallel lines. _____

3 Identify two intersecting lines that are not perpendicular.

_____

Classify each angle as straight, right, obtuse, or acute. Then, use a protractor to measure each angle. You may need to extend its sides.

4

5

6

_____   _____   _____

7 Place a check mark under the equilateral triangle.

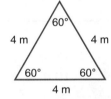 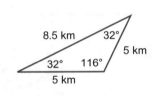 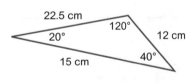 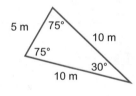

___   ___   ___   ___

8 Which shape is not a polygon? Place a check mark under it.

___   ___   ___   ___

9 Place a check mark under the trapezoid.

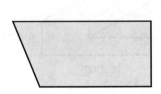 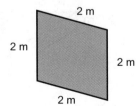 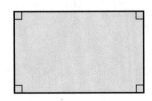

___   ___   ___   ___

Name _____

Find the area of each shape.

10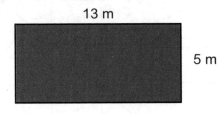

13 m

5 m

_____

11 a square with one side measuring 17 cm

_____

Find the volume of each object.

12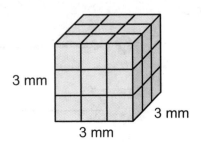

3 mm

3 mm

3 mm

_____

15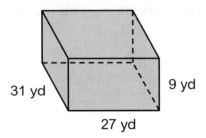

31 yd

9 yd

27 yd

_____

13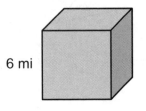

6 mi

_____

16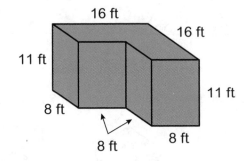

16 ft

16 ft

11 ft

11 ft

8 ft

8 ft

8 ft

_____

14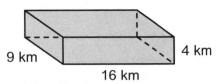

9 km

4 km

16 km

_____

17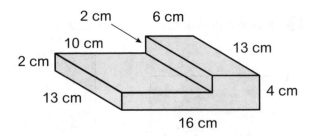

2 cm

6 cm

10 cm

13 cm

2 cm

13 cm

4 cm

16 cm

_____

Making a Line Plot

Data is a collection of numbers or pieces of information. A line plot is a diagram that uses a number line to show data.

Example

This line plot shows the length in meters of seven hammerhead sharks. Each X shows one number from the data set.

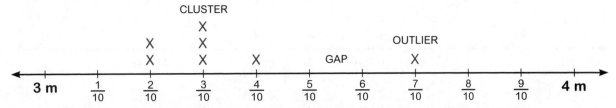

The line plot shows that three sharks are $3\frac{3}{10}$ meters long. It shows that two sharks are $3\frac{2}{10}$ meters long, and one shark is $3\frac{4}{10}$ meters long. The longest shark is $3\frac{7}{10}$ meters long.

Several data points within a small interval form a **cluster**. An interval that contains no data points is a **gap**. Any number that is much greater or much less than other data in the set is an **outlier**.

Diagram and Interpret

The list of data shows the mass in kilograms of nine cats.

Harry, $3\frac{3}{4}$ kg   Sophie, $3\frac{1}{4}$ kg   Grace, $3\frac{1}{2}$ kg

Ernie, $3\frac{1}{2}$ kg   Molly, 4 kg   Bob, $3\frac{1}{4}$ kg

Squeaker, $2\frac{1}{4}$ kg   Mercedes, $4\frac{1}{4}$ kg   Blackie, $3\frac{1}{2}$ kg

1 Draw a line plot to show this data set. Use the line plot for exercises 2–5.

2 Which mass is an outlier?

_____

3 Draw a rectangle around the cluster. Draw an oval around the gap.

4 How many cats have a mass of $3\frac{1}{4}$ kg?

_____

5 What is the most common mass in the data set?

_____

Name _____

Using Line Plots to Solve Problems

Example

Students in a gardening club grew some tomato plants one summer. This line plot shows the height of their tomato plants.

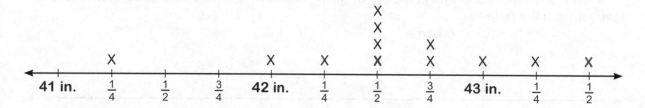

Gavin's plant is shown with the red X. Greta's plant is shown with the blue X. What is the difference in height between Gavin's plant and Greta's plant?

Step 1: Read the line plot to find the information you need:

The height of Gavin's plant: $42\frac{1}{2}$ in.

The height of Greta's plant: $43\frac{1}{2}$ in.

Step 2: Decide which operation you should use to solve. In this case, you need to find a difference. You subtract.

$$43\frac{1}{2} - 42\frac{1}{2} = 1$$

The difference in height between Gavin's plant and Greta's plant is 1 in.

Solve

Use the line plot above for exercises 1–6.

1 What is the difference between the tallest plant and the shortest plant shown on the line plot?

_____ $2\frac{1}{4}$ in. _____

2 What is the sum of the height of the two tallest plants?

_____

3 How many plants are $42\frac{1}{2}$ in. tall? _____

What is the total sum of their height?

_____

4 What is the total height of the three shortest plants?

_____

5 Which height is an outlier?

_____

6 To find the average height, you would find the sum of all the heights shown and then divide by the number of plants. What is the average height shown on the line plot?

_____

Plotting Points on a Coordinate Grid

You can locate points on a map using a coordinate grid.

Example

A coordinate grid has a horizontal *x*-axis. It also has a vertical *y*-axis. The point at which the axes intersect is called the **origin**.

An **ordered pair** names a point on the grid. In the coordinate grid at the right: the store is at (4,3); home is at (2,5).

The *x*-coordinate is the first number in an ordered pair. It names the distance to the left or right of the origin along the *x*-axis. The *y*-coordinate is the second number in an ordered pair. It names the distance up or down from the origin along the *y*-axis.

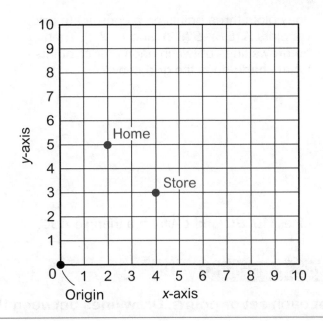

Plot and Label

Plot and label each point on the grid. The first point has been plotted and labeled for you.

1. A (2,3)

2. B (7,8)

3. C (4,5)

4. D (2,9)

5. E (3,1)

6. F (8,4)

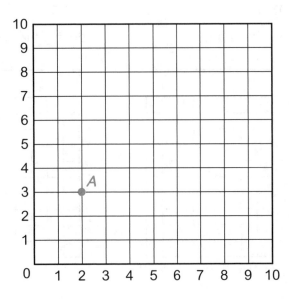

Name _____

Plotting Points to Form Lines and Shapes

You can show lines and shapes on a coordinate grid by plotting points.

Example

Joey plots three points on a coordinate grid. He plots A (5,8), B (8,2), and C (2,2). Then he draws lines between each pair of points. What shape does the grid show?

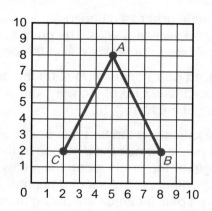

Lines $\overline{AB}$, $\overline{BC}$, and $\overline{CA}$ form a triangle ABC.

Plot and Draw

Plot each set of points. Draw lines between the points. Identify the shape.

1 A (3,8)

B (7,8)

C (7,4)

D (3,4)

Identify the shape.

_____square_____

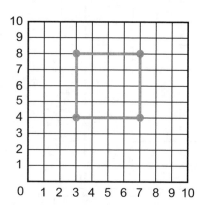

2 A (3,8)

B (7,8)

C (9,5)

D (1,5)

Identify the shape.

_____

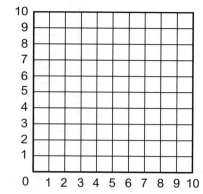

3 R (3,6)

S (8,5)

T (2,3)

Identify the shape.

_____

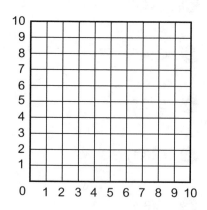

Analyzing Shapes and Lines

Kelly draws $\overline{AB}$ on the grid below. She wants to draw more lines to show a square. How can she draw a square on the coordinate grid?

Remember! Intersecting lines cross each other through a single point. Perpendicular lines form square corners when they cross. Parallel lines never cross and are always the same distance apart.

Example

Step 1: Count the number of squares between point A and point B.
Step 2: Draw a line through B that is perpendicular to $\overline{AB}$. Count the same number of squares from point B, and draw a new point. Label it C.
Step 3: Draw a line through A that is perpendicular to $\overline{AB}$. Count the same number of squares from point A, and draw a new point. Label it D.
Step 4: Draw a line through points C and D.

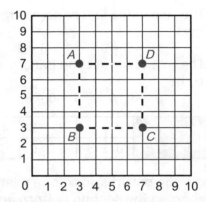

Plot and Draw

1 On the grid, draw lines $\overline{AB}$, $\overline{BC}$, and $\overline{CA}$. What shape did you draw?

_____ triangle _____

What are the coordinates of the shape?

_____

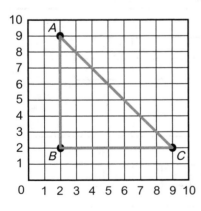

2 Draw a rectangle on the grid. Be sure each corner is a point. Label the points.

What are the coordinates of the points you drew?

_____

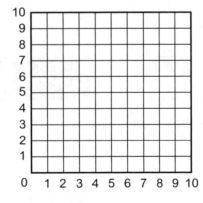

3 Kim is building a flower box. She draws lines $\overline{AB}$, $\overline{CD}$, and $\overline{AC}$. She wants to draw a fourth side parallel to $\overline{AC}$. Draw the line.

What shape is the flower box?

_____

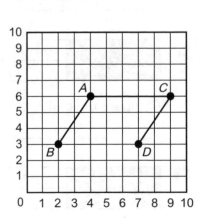

Name _____

Creating Number Patterns

You can create number patterns if you know the pattern rules.

Example

How can you complete the patterns shown?

Multiply by 3	0	1	2	3	4
	0	3			

Multiply by 6	0	1	2	3	4

Step 1: The rule of the first pattern is multiply by 3. To complete the table, multiply each number by 3.

$0 \times 3 = 0$ $1 \times 3 = 3$ $2 \times 3 = 6$ $3 \times 3 = 9$ $4 \times 3 = 12$

Step 2: The rule of the second pattern is multiply by 6. To complete the table, multiply each number by 6.

$0 \times 6 = 0$ $1 \times 6 = 6$ $2 \times 6 = 12$ $3 \times 6 = 18$ $4 \times 6 = 24$

Multiply by 3	0	1	2	3	4
	0	3	6	9	12

Multiply by 6	0	1	2	3	4
	0	6	12	18	24

To compare the patterns, look for a relationship between the two patterns:
The numbers in the second pattern are twice the numbers in the first pattern.
Zero is the only exception.

Calculate

Complete the patterns in each set of tables. Then compare each pattern and write about your comparison.

1

Add 4	0	1	2	3	4
	4	5	6	7	8

Add 7	0	1	2	3	4
	7	8	9	10	11

Compare the patterns. What is their relationship?

Each number in the second pattern is 3 more than the number in the first pattern.

2

Subtract 2	10	9	8	7	6

Subtract 5	10	9	8	7	6

Compare the patterns. What is their relationship?

_____

3

Multiply by 4	0	1	2	3	4

Multiply by 2	0	1	2	3	4

Compare the patterns. What is their relationship?

_____

Plotting Patterns on Coordinate Grids

The numbers in patterns can be ordered pairs. You can also plot them on coordinate grids.

Example

Marsha has two charts that show patterns. How can she plot them on a coordinate grid?

Multiply by 2	0	1	2	3	4
	0	2	4	6	8

Multiply by 1	0	1	2	3	4
	0	1	2	3	4

Step 1: Plot the first set of numbers from the first pattern. The ordered pairs are (0,0), (1,2), (2,4), (3,6), (4,8). Draw a line through the points.

Step 2: Plot the second set of numbers from the second pattern. The ordered pairs are (0,0), (1,1), (2,2), (3,3), (4,4). Draw a line through the points. Use a different color to draw the line so it is easier to see.

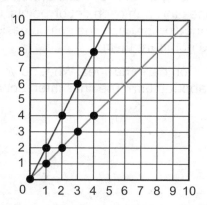

Calculate and Plot

Complete the patterns in each set of tables. Then plot each set of patterns in the coordinate grid.

1

Add 2	0	1	2	3	4
	2	3	4	5	6

Add 3	0	1	2	3	4
	3	4	5	6	7

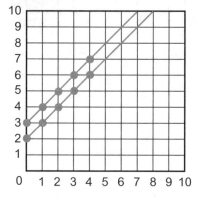

2

Subtract 2	9	8	7	6	5

Subtract 1	9	8	7	6	5

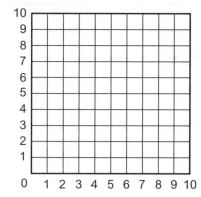

3

Multiply by 2	1	2	3	4	5

Add 2	1	2	3	4	5

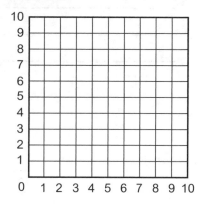

Name _____

Using Grids to Compare Patterns

You can use grids to compare and explain relationships among patterns.

Example

How can you use a grid to compare the relationships of these patterns?

Multiply by 2	0	1	2	3	4
	0	2	4	6	8

Add 3	0	1	2	3	4
	3	4	5	6	7

Step 1: Plot both sets of numbers on a grid. Draw lines through each set of points. Use a different color for each line so it is easier to see and compare them.

Step 2: Compare the lines. What do you notice about them? The lines intersect at (3,6).

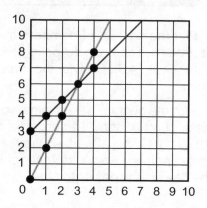

Plot and Compare

Plot the patterns on the coordinate grid. Then compare the lines. Describe your comparison.

Add 4	0	1	2	3	4
	4	5	6	7	8

Multiply by 3	0	1	2	3
	0	3	6	9

Compare the lines on the grid. Describe your comparison.

_____

_____

_____

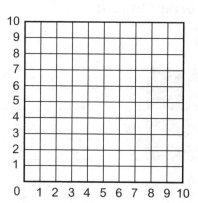

This line plot shows the height in meters of seven young orange trees.

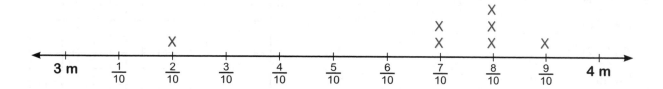

1 Which height is an outlier?

_____

2 How many trees have a height of $3\frac{7}{10}$ m?

_____

3 What is the most common height in the data set?

_____

4 What is the total height of the three shortest trees?

_____

5 What is the difference between the tallest tree and the shortest tree shown on the line plot?

_____

6 To find the average height, you would find the sum of all the heights shown and then divide by the number of trees. What is the average height shown on the line plot?

_____

Plot and label each point on the coordinate grid.

7 A (2,3)

8 B (6,4)

9 C (4,8)

10 D (7,13)

11 E (10,10)

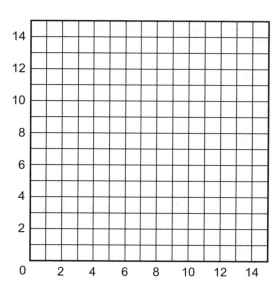

Name _____

Plot each point. Draw lines between the points. Identify the shape.

12 A (8,11)

B (2,2)

C (14,2)

Identify the shape.

_____

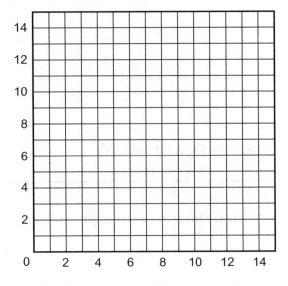

Plot each point. Draw lines between the points. Identify the shape.

13 A (2,4)

B (2,10)

C (8,10)

D (8,4)

Identify the shape.

_____

Name _____

Write each number in standard form.

1 39 thousandths

2 17 and 871 thousandths

3 84 billion, 652 million, 34

_____   _____   _____

Write each number in word form.

4 7,111,658,314

_____

5 12,500,500,500

_____

Write >, <, or = to compare each pair of numbers.

6 1.1 ____ 0.99

7 0.012 ____ 0.121

8 5.687 ____ 5.687

Round each number to the place of the underlined digit.

9 7.1<u>2</u>3 _____

10 <u>4</u>.741 _____

11 6.<u>1</u>48 _____

Add. Write the sum.

12
```
  3.87
+ 2.66
```

13
```
  791
+ 109
```

14
```
  680
+  64
```

15
```
  4.72
+ 0.16
```

16 61,115 + 17,956 _____

17 8.46 + 0.54 _____

18 67.36 + 1.51 _____

Subtract. Write the difference.

19
```
  411
-  24
```

20
```
  7.01
- 6.21
```

21
```
  815
- 656
```

22
```
  214.14
-  37.09
```

23 77,777 − 3,028 = _____

24 12,499 − 11,645 = _____

25 17.66 − 14.76 = _____

26 An alligator's brain weighs about 0.49 ounces. A peach can weigh about 5.5 ounces. How many ounces heavier is a peach than an alligator's brain?

_____

Simplify and complete. Tell what property is represented.

27 $15 \times 12 = ($ _____ $\times 12) + (10 \times$ _____ $) = 60 +$ _____ $=$ _____

_____ Property

28 $(8 \times 2) \times 30 =$ _____ $\times (2 \times$ _____ $) = 8 \times$ _____ $=$ _____

_____ Property

Multiply. Write the product.

29 $\begin{array}{r} 256 \\ \times\ \ \ 3 \\ \hline \end{array}$
  30 $\begin{array}{r} 14 \\ \times\ \ 9 \\ \hline \end{array}$
  31 $\begin{array}{r} 57 \\ \times\ 57 \\ \hline \end{array}$
  32 $\begin{array}{r} 764 \\ \times\ 12 \\ \hline \end{array}$

33 A gym has 2-dozen treadmills. Each treadmill weighs 145 pounds. How many pounds do the treadmills weigh in all?

_____

34 $1{,}685 \times 10^2 =$ _____   **35** $59 \times 10^4 =$ _____   **36** $987 \times 10^1 =$ _____

Divide. Write the quotient.

37 $28 \div 5 =$ _____   **38** $90 \div 2 =$ _____   **39** $77 \div 8 =$ _____

Estimate. Then multiply or divide.

40 Estimate: $3.02 \times 4 =$ _____

Multiply: $3.02 \times 4 =$ _____

41 Estimate: $4.95 \times 6 =$ _____

Multiply: $4.95 \times 6 =$ _____

42 Estimate: $6.14 \times 3 =$ _____

Multiply: $6.14 \times 3 =$ _____

43 $15.88 \div 10^2$

Estimate: _____

Quotient: _____

44 $872.3 \div 10^4 =$

Estimate: _____

Quotient: _____

45 $298.4 \div 10^1 =$

Estimate: _____

Quotient: _____

46 $1.37 \times 10^2 =$

Estimate: _____

Product: _____

47 $7.743 \times 10^4 =$

Estimate: _____

Product: _____

48 $15.85 \times 10^3 =$

Estimate: _____

Product: _____

Work backward to solve.

49 Mr. Pedersen is making a pot of marinara sauce. He pours the sauce into 8 jars. Each jar holds 1.5 quarts of sauce. There are 0.75 quarts of sauce left in the pot. How many quarts of sauce were in the pot before Mr. Pedersen began pouring?

_____

Name _____

Simplify and solve. Show your work.

50 $(18 - 8)^3 - (12 \times 5) \times (9 - 4) + 10^4$

Simplify inside the parentheses: _____

Simplify exponents: _____

Multiply: _____

Add and subtract from left to right: _____

Final answer: _____

51 $6\{[4(1 + 7) + 3] - 15\}$

Simplify inside the parentheses: _____

Simplify inside the brackets: _____

Simplify inside the braces: _____

$6\{[4(1 + 7) + 3] - 15\} =$ _____

52 $(72 \div 8) + (13 - 3)^2$

53 $14 + 4^2 \times (2 \times 4) \div 2$

_____

_____

Add or subtract. Give the answers in simplest terms. Change any improper fractions to mixed numbers.

54 $\frac{1}{6} + \frac{2}{6}$ _____

55 $\frac{3}{9} + \frac{1}{9}$ _____

56 $\frac{7}{8} - \frac{1}{8}$ _____

57 $\frac{4}{9} + \frac{5}{6}$ _____

58 $\frac{1}{4} - \frac{1}{8}$ _____

59 $\frac{5}{7} - \frac{2}{9}$ _____

60 $\quad 9\frac{3}{8}$
$\quad - 2\frac{1}{5}$

61 $\quad 7\frac{2}{3}$
$\quad + 4\frac{3}{8}$

62 $\quad 2\frac{5}{6}$
$\quad + 4\frac{1}{2}$

63 $\quad 5\frac{2}{7}$
$\quad - 3\frac{3}{4}$

Find the equivalent fraction.

64 equivalent to $\frac{2}{3}$, denominator 9 _____

66 equivalent to $\frac{1}{4}$, denominator 28 _____

65 equivalent to $\frac{1}{2}$, denominator 20 _____

67 equivalent to $\frac{7}{8}$, denominator 16 _____

Multiply.

68 $24 \times \frac{1}{6}$ _____

69 $\frac{5}{8} \times 21$ _____

70 $7 \times \frac{1}{7}$ _____

Solve.

71 A florist is making an arrangement of 45 roses. The customer has asked that $\frac{2}{3}$ of the roses be pink. How many pink roses does the florist need?

_____

72 $\frac{3}{4}$ of Diego's grocery cart is filled with fresh produce. $\frac{6}{7}$ of the produce is leafy greens. What fraction of the produce is leafy greens?

_____

73 Chase's journal has 200 pages. So far, he has written 32 pages of journal entries. If he writes $\frac{1}{2}$ a page every day, how many days will it take him to fill the remaining pages?

_____

Multiply to find the area of each rectangle.

74 9 ft long $\times \frac{3}{8}$ ft wide

area = _____

75 5 m long $\times \frac{1}{3}$ m wide

area = _____

Find each quotient. Draw models to help. Use multiplication to check your answers.

76 $\frac{1}{6} \div 9 =$ _____

77 $\frac{1}{2} \div 7 =$ _____

78 $9 \div \frac{1}{8} =$ _____

Solve.

79 The reptile exhibit at a zoo has a python that is 5 meters long. There is also a coral snake that is 0.6 meters long. How much longer in centimeters is the python than the coral snake?

_____

80 People must be at least 54 inches tall to ride a roller coaster at an amusement park. Gabrielle is 4 feet 3 inches tall. Is she tall enough to ride the coaster? How do you know?

_____

81 Yolanda is 1 year, 5 days old. Her brother Benjamin is 4 years, 1 week old. How many days older is Benjamin than Yolanda?

_____

82 What temperature does the thermometer show? _____

The average temperature in Paris in July is 75°F. Is the temperature shown by the thermometer higher or lower than average? How many degrees higher or lower?

_____

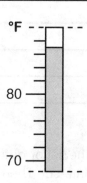

Use the diagram for Exercises 83 to 85.

83 Identify 3 points. _____

84 Identify 2 intersecting lines that are not perpendicular. _____

85 Identify 3 line segments. _____

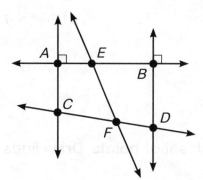

Classify each angle as straight, right, obtuse, or acute. If you have a protractor, measure each angle. You may need to extend the angle to measure it. Write each measure.

86

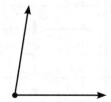

_____

87

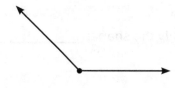

_____

Name _____

Place a check mark below the figures that are not polygons.

88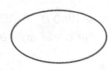

_____ _____ _____ _____

89 Anthony says that trapezoids are parallelograms because trapezoids have a pair of parallel sides. Do you agree? Why or why not?

_____

_____

_____

_____

90 Find the volume of the object.

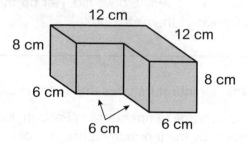

_____

91 The list of data shows the mass in kilograms of 6 flamingos at a zoo. Draw a line plot to show this data set. Then draw a circle around the outlier.

Pinky, $1\frac{1}{2}$ kg   Olaf, $3\frac{1}{4}$ kg   Neena, $3\frac{1}{2}$ kg

Marcus, 3 kg   Hunter, $3\frac{1}{2}$ kg   Olivia, $3\frac{1}{2}$ kg

Plot each set of points. Draw lines between the points. Identify the shape.

92 A (3,2)

B (5,2)

C (3,7)

D (5,7)

Identify the shape: _____

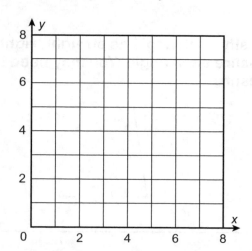

Acute Angle: An angle that is less than 90 degrees. *(p. 119)*

Acute Triangle: A triangle with three acute angles. *(p. 120)*

Angle: A figure formed by two rays that share the same endpoint. *(p. 119)*

Area: The number of square units needed to cover a region. *(p. 124)*

Array: A way to show equal groups of objects in columns and rows. *(p. 41)*

Associative Property of Addition: You can change the grouping of addends. The sum remains the same. *(p. 23)*

Associative Property of Multiplication: You can change the grouping of factors. The product remains the same. *(p. 37)*

Base: The number to the left of the exponent that is multiplied by itself. *(p. 16)*

10 is the base in 10^2.

Braces { }: Symbols used to group numbers or letters. The work in braces is done after the work in brackets and parentheses. *(p. 63)*

$$3\{2 + [2(1 + 1)]\} =$$
$$3\{2 + [2(2)]\} = 3\{2 + [4]\} =$$
$$3\{6\} = 18$$

Brackets [ ]: Symbols used to group numbers or letters. The work in brackets is done after the work in parentheses. *(p. 63)*

$$[3(1+1) + 4] = [3(2) + 4] =$$
$$[6 + 4] = 10$$

Capacity: The amount a container can hold measured in liquid units. *(p. 107)*

Celsius (°C): A metric unit used to measure temperature. *(p. 109)*

Centimeter: A metric unit used to measure length. *(p. 105)*

Cluster: Data points that are grouped closely together on a line plot. *(p. 133)*

Commutative Property of Addition: You can add numbers in any order. The sum remains the same. *(p. 23)*

Commutative Property of Multiplication: You can multiply factors in any order. The product remains the same. *(p. 36)*

Coordinate Grid: A grid with numbered lines used to locate specific points. *(p. 135)*

Cubic Unit: The volume of a cube that measures 1 unit high, 1 unit wide, and 1 unit long. *(p. 125)*

Cup: A customary unit of capacity. *(p. 113)*

Customary System: A measurement system that uses units such as inches and feet. *(p. 111)*

Data: Gathered information. *(p. 133)*

Deciliter: A metric unit used to measure capacity. *(p. 107)*

Decimal Number: A number with one or more digits to the right of a decimal point. *(p. 17)*

Decimal Point: A dot used to separate ones from tenths in a decimal number. *(p. 17)*

Degree (°): A unit of measurement for angles. *(p. 119)*

Denominator: The total number of equal parts of a whole or a set. The number below the fraction bar in a fraction. *(p. 68)*

Digit: The symbols 0, 1, 2, 3, 4, 5, 6, 7, 8, and 9 that are used to write numbers. *(p. 17)*

Distributive Property of Multiplication: You can multiply two addends by a factor. You can also multiply each addend by the same factor and add the products. The total is the same. *(p. 37)*

Dividend: The number divided in a division problem. *(p. 46)*

Divisor: The number of equal groups a dividend is divided by in a division problem. *(p. 46)*

Endpoint: A point at the beginning of a ray or at either side of a line segment. *(p. 118)*

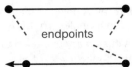

endpoints

Equation: A number sentence that uses an equal sign (=). *(p. 57)*

Equilateral Triangle: A triangle in which all sides are the same length. *(p. 120)*

Equivalent Fractions: Fractions that name the same part of a whole. *(p. 69)*

$$\frac{1}{2} = \frac{3}{6}$$

Estimate: To find a number close to an exact amount. To make a careful guess. *(p. 76)*

Expanded Form: A number form that shows the value of the digits in a number. *(p. 14)*

$$437 = 400 + 30 + 7$$

Exponent: A number that shows how many times the base is multiplied by itself. *(p. 16)*

$10^2 = 10 \times 10$; 2 is the exponent

Factors: The numbers you multiply in a multiplication problem. *(p. 89)*

Fahrenheit (°F): A customary unit used to measure temperature. *(p. 110)*

Foot, Feet: A customary unit used to measure length. *(p. 111)*

Fraction: Symbols, such as $\frac{1}{2}$ or $\frac{2}{3}$, used to name part of a whole or parts of a set. *(p. 68)*

Front-End Estimation: Estimating to the first digit in a number. *(p. 76)*

587 becomes 500

Gallon: A customary unit used to measure capacity. *(p. 113)*

Gap: Large spaces between data points on a line plot. *(p. 133)*

Gram: A metric unit used to measure mass. *(p. 106)*

Hexagon: A polygon that has six sides. *(p. 122)*

Hundredth: One part of 100 equal parts. *(p. 17)*

Identity Property of Multiplication: Any number multiplied by 1 equals that number. *(p. 36)*

Improper Fraction: A fraction with a numerator greater than or equal to the denominator. *(p. 72)*

Inch: A customary unit used to measure length. *(p. 111)*

Intersecting Lines: Lines that cross at a single point. *(p. 118)*

Inverse Operations: Operations that undo each other. *(p. 97)*

Irregular Polygon: A polygon with angles of unequal measure and sides of unequal length. *(p. 123)*

Isosceles Triangle: A triangle that has two sides of equal length. *(p. 120)*

Kilogram: A metric unit used to measure mass. *(p. 106)*

Kilometer: A metric unit used to measure length. *(p. 105)*

Least Common Multiple (LCM): The least, or smallest, number that is a multiple of two numbers. *(p. 68)*

Line: A straight path through two points. It goes on without end in both directions. *(p. 118)*

Line Plot: A way to show data with a number line. *(p. 133)*

Line Segment: Part of a line with two endpoints. *(p. 118)*

Liter: A metric unit used to measure capacity. *(p. 107)*

Glossary

Mass: The measure of the amount of matter in an object. *(p. 106)*

Metric System: A measurement system that uses units such as meters and liters. *(p. 105)*

Meter: A metric unit used to measure length. *(p. 105)*

Metric Ton: A metric unit used to measure mass. *(p. 106)*

Mile: A customary unit used to measure length. *(p. 111)*

Milligram: A metric unit used to measure mass. *(p. 106)*

Milliliter: A metric unit used to measure capacity. *(p. 107)*

Mixed Number: A number containing a whole number and a fraction. *(p. 72)*

Numerator: A specific number of equal parts of a whole or a set. The number above the fraction bar in a fraction. *(p. 69)*

Obtuse Angle: An angle that is more than 90 degrees. *(p. 119)*

Obtuse Triangle: A triangle with one obtuse angle. *(p. 120)*

Octagon: A polygon that has eight sides. *(p. 122)*

Ordered Pair: A pair of numbers that locates a point on a coordinate grid. *(p. 135)*

Order of Operations: Rules about the order in which the steps in a problem are carried out. *(p. 61)*

Origin: The point where two axes intersect on a coordinate grid. *(p. 135)*

Ounce: A customary unit of weight. *(p. 112)*

Outlier: Any number that is very different from the rest of the numbers in a data set. *(p. 133)*

Parallel Lines: Lines that are always the same distance apart and never cross. *(p. 118)*

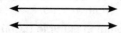

Parallelogram: A four-sided figure in which opposite sides are parallel *(p. 121)*

Parentheses ( ): Symbols used to group numbers or letters. *(p. 61)*
$$(3 \times 1) + 1 = 3 + 1 = 4$$

Partial Products: Products found by breaking a factor into ones, tens, hundreds, and so on and then multiplying each of these by another factor. *(p. 41)*

Pentagon: A polygon that has five sides. *(p. 122)*

Perpendicular Lines: Lines that form right angles when they cross. *(p. 118)*

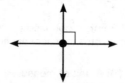

Pint: A customary unit used to measure capacity. *(p. 113)*

Place Value: The value of a digit. *Example:* The value of 4 in 5,429 is 400. *(p. 14)*

Plane: An endless flat surface. *(p. 122)*

Point: An exact location in space. *(p. 135)*

A

Polygon: A closed plane figure. *(p. 122)*

Pound: A customary unit used to measure weight. *(p. 112)*

Prime Number: A whole number that has only two factors, 1 and itself. *(p. 68)*

Product: The answer to a multiplication problem. *(p. 36)*

Protractor: A math tool used to measure angles. *(p. 119)*

Quadrilateral: A polygon that has four sides and four angles. *(p. 121)*

Quart: A customary unit used to measure capacity. *(p. 113)*

Quotient: The answer to a division problem. *(p. 46)*

Ray: A part of a line with one endpoint. It goes on without end in the other direction. *(p. 118)*

Regular Polygon: A polygon with angles of equal measure and sides of equal length. *(p. 123)*

Remainder: The amount left over after division is complete. *(p. 45)*

Rhombus: A four-sided figure in which opposite sides are parallel and all four sides are equal length. *(p. 121)*

Right Angle: An angle that measures exactly 90 degrees. *(p. 119)*

Right Triangle: A triangle with one right angle. *(p. 120)*

Round, Rounding: To express a number to the nearest ten, hundred, thousand, and so on. *(p. 19)*

Scalene Triangle: A triangle in which no sides are of equal length. *(p. 120)*

Shortened Word Form: A way to write a number using digits and words. *(p. 14)*
6 billion, 100 million, 315

Standard Form: A way to write a number using only digits. *(p. 14)*
2,350

Straight Angle: An angle that measures exactly 180 degrees. *(p. 119)*

Sum: The answer to an addition problem. *(p. 24)*

Tenth: One out of 10 equal parts. *(p. 17)*

Thousandth: One out of 1,000 equal parts. *(p. 17)*

Ton: A customary unit used to measure weight. *(p. 112)*

Trapezoid: A four-sided figure in which only one pair of sides is parallel. *(p. 121)*

Unit Fraction: A fraction with 1 as the numerator. *(p. 96)*

Variable: A symbol that stands for a number. *(p. 115)*

Vertex: The endpoint that two rays share to form an angle. *(p. 119)*

Volume: The number of cubic units that fits inside a solid figure. *(p. 125)*

Weight: The measure of how heavy something is. *(p. 112)*

Word Form: A way to write a number using words. *(p. 14)*
four hundred seventy-eight

x-axis: A horizontal numbered line on a coordinate grid. *(p. 135)*

x-coordinate: The first number in an ordered pair. *(p. 135)*

Yard: A customary unit used to measure length. *(p. 111)*

y-axis: A vertical numbered line on a coordinate grid. *(p. 135)*

y-coordinate: The second number in an ordered pair. *(p. 135)*

Pretest
Pages 8 – 13
1. 53.125   2. .096
3. 2,109,000,007
4. nineteen billion, one hundred three million, five hundred seventy-eight thousand, one hundred forty-three
5. two hundred million, two hundred thousand, two hundred
6. <   7. <
8. =   9. 4.37
10. 1.03   11. 17
12. 9.53   13. 1,180
14. 801   15. 6.3
16. 72,807   17. 20,410
18. 77.93   19. 624
20. 5.91   21. 93
22. 277.71   23. 1,714
24. 108.25   25. 4.2
26. 2.69 ounces
27. 10; 8; 80; 160; Distributive Property
28. 30; 5; 20; 600; Associative Property
29. 138   30. 392
31. 2,044   32. 9,225
33. $120   34. 281,100
35. 61,000   36. 776
37. 27   38. 3 R2
39. 13   40. > 12; 12.4
41. > 35; 35.65
42. > 66; 66.54
43. >.0019; .00192
44. > 8; 8.723
45. > .016; 0166
46. < 15,000; 14,850
47. > 10,000; 12,560
48. < 7,000; 6,798
49. $2^3 + 24 \times 2 + 10^3$; $8 + 24 \times 2 + 1000$; $8 + 48 + 1000$; 1056 1056
50. $4\{[2(7) + 2] - 9]\}$; $4\{16 - 9\}$; $4\{7\}$; 28
51. 14   52. 28
53. $\frac{3}{4}$ 54. $\frac{7}{8}$
55. $\frac{2}{3}$ 56. $3\frac{1}{12}$
57. $\frac{13}{24}$ 58. $\frac{25}{18} = 1\frac{7}{18}$
59. $1\frac{3}{35}$ 60. $10\frac{1}{6}$

61. $15\frac{7}{10}$ 62. $7\frac{5}{6}$
63. $\frac{16}{24}$ 64. $\frac{21}{42}$
65. $\frac{12}{36}$ 66. $\frac{12}{32}$
67. $\frac{80}{8} = 10$ 68. $\frac{36}{3} = 12$
69. $\frac{12}{50} = \frac{6}{25}$ 70. 6 inches
71. $\frac{1}{10}$
72. 80 − 20 = 60; The mother is about 60 feet longer than the baby.
73. $\frac{27}{4} = 6\frac{3}{4}$ sq ft
74. $\frac{8}{2} = 4$ sq m
75. $\frac{1}{36}$; $\frac{1}{36} \times 9 = \frac{9}{36} = \frac{1}{4}$
76. $\frac{1}{10}$; $\frac{1}{10} \times 5 = \frac{5}{10} = \frac{1}{2}$
77. 96; $96 \times \frac{1}{8} = \frac{96}{8} = 12$
78. 450 mm   79. 230 grams
80. 700 mL
81. 6 minutes
82. 53 inches 83. obtuse
84. right   85. straight
86. 12   87. 16
88. 18
89. equilateral triangle; acute triangle
90. isosceles triangle, obtuse triangle
91.

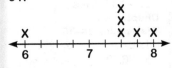

92. 2 ft   93. $22\frac{1}{2}$ ft
94. (4,4)   95. (1,1)
96. (2,3)   97. (3,2)

Chapter 1

Chapter 1
Lesson 1, page 14
1. (2 × 1,000,000,000) + (7 × 100,000,000) + (1 × 10,000,000) + (5 × 1,000,000) + (4 × 100,000) + (3 × 10,000) + (4 × 1,000) + (1 × 100) + (2 × 10) + (1 × 1)

2. (5 × 1,000,000,000) + (4 × 10,000,000) + (2 × 1,000,000) + (6 × 100,000) + (9 × 10,000) + (9 × 1,000) + (1 × 1)
3. four billion, two hundred fifty-seven million, seven hundred ninety-three thousand, four hundred thirty-six
4. six billion, two hundred seven million, four hundred ninety thousand, twenty-one
5. 1 billion, 450 million, 212
6. 5 billion, 842 million, 776 thousand, 481
Chapter 1
Lesson 2, page 15
1. 80 or 8 tens; 8 or 8 ones; Possible answer: It is 10 times greater.
2. 7 or 7 ones; 70 or 7 tens; Possible answer: It has $\frac{1}{10}$ the value.
3. Possible answer: It is ten times greater.
4. Possible answer: It has $\frac{1}{10}$ the value.
5. 753 and 735
Chapter 1
Lesson 3, page 16
1. 75,600 × (10 × 10) = 75,600 × 100 = 7,560,000
2. 75,600 ÷ (10 × 10) = 75,600 ÷ 100 = 756
3. 4 × (10 × 10 × 10 × 10) = 4 × 10,000 = 40,000
4. 4 ÷ (10 × 10 × 10 × 10) = 4 ÷ 10,000 = $\frac{4}{10,000}$ or 0.0004
5. 23 ÷ (10 × 10 × 10) = 23 ÷ 1,000 = $\frac{23}{1,000}$ or 0.023
6. 95,500   7. 53,000
8. Possible answer: The decimal point moves to the right when I multiply by a power of 10. If I divide by a power of 10, the decimal point moves to the left.
Chapter 1
Lesson 4, page 17
1. (3 × 10) + (5 × 1) + (7 × $\frac{1}{10}$) + (2 × 1/100) + (6 × $\frac{1}{1,000}$)

2. (6 × 100) + (9 × 10) + (9 × 1) + (4 × $\frac{1}{1,000}$)
3. forty-two and five hundred seventy-six thousandths
4. 5.401   5. 0.024
6. 1 and 451 thousandths
7. 1 and 1 thousandth
Chapter 1
Lesson 5, page 18
1. <   2. >
3. <
4. 0.345, 0.32, 0.305
5. 7.29, 0.729, 0.079
6. 0.456, 0.45, 0.4
7. half-dollar, nickel, quarter, cent, dime
Chapter 1
Lesson 6, page 19
1. 6.5

6.5   6.55   6.6

2. 0.74

0.74   0.745   0.75

3. 0.35   4. 7.3
5. 3   6. 3.56
7. 0.99   8. 9.1
9. 120.3 g   10. .04 mm
Chapter 1
Lesson 7, page 20
1. Rafael, Planetarium; Maria, Natural History Museum; Jose, Aquarium
2. Alex, skiing; Brian, skateboarding; Candy, rollerblading; Diane, speed skating
Chapter 1
Test, pages 21–22
1. 1 billion, four hundred thirty-two thousand, five hundred and six
2. 200 or 2 hundreds; 20 or 2 tens; Possible answer: It is 10 times greater.
3. Possible answer: It has $\frac{1}{10}$ the value.
4. 23 × (10 × 10 × 10); 23,000; 23 thousand
5. 640,000   6. 3.03
7. 5,000   8. 65,350
9. 0.043   10. 77
11. 6 and 92 hundredths
12. 875 thousandths
13. 72 and 7 tenths
14. >   15. <
16. <   17. =
18. >   19. <
20. 23.3   21. 2.4

Answers

22. 1.1   **23.** 23.27
24. 2.45   **25.** 1.12
26. 57,000,000,000
27. 100 million; Possible answer: It has 10 times the value
28. 80; Possible answer: It has $\frac{1}{10}$ the value
29. $(2 \times 10) + (3 \times 1) + (3 \times \frac{1}{10}) + (9 \times \frac{1}{1000})$
30. 1 and 812 thousandths
31. sixty-two and one hundred fifty-nine thousandths
32. 181.30
33. Charlotte, Birmingham, Atlanta, Denver

Chapter 2

Chapter 2
Lesson 1, page 23
1. 127; 127   **2.** 125; 125
3. 131; 131   **4.** 153; 153
5. 169   **6.** 145
7. 128   **8.** 142
9. 171   **10.** 156
11. Possible answer: It would be easiest to add 60 to 26 first. She can add those numbers in her head to get 86. The Commutative Property states that you can add in any order, and the sum, in this case 143, will be the same.

Chapter 2
Lesson 2, page 24
1. 17   **2.** 39
3. 147; 64   **4.** 56
5. 79   **6.** 24; 82
7. 71   **8.** 79
9. 134; 78
10. Answers will vary. Teddy can subtract 49 from 56 to find the value of x,
11. Teresa can subtract 117 from 136 to find the difference.

Chapter 2
Lesson 3, page 25
1. 289   **2.** 484
3. 842   **4.** 449
5. 345   **6.** 990
7. 504   **8.** 746
9. 201   **10.** 830
11. 418   **12.** 633
13. 2,290   **14.** 59,890

15. 4,343   **16.** 79,022
17. 229,463

Chapter 2
Lesson 4, page 26
1. 232   **2.** 110
3. 131   **4.** 308
5. 132   **6.** 50
7. 285   **8.** 270
9. 768   **10.** 304
11. 623   **12.** 26
13. 2,019   **14.** 22,103
15. 4,229   **16.** 48,845
17. 4,877 kg

Chapter 2
Lesson 5, page 27
1. 2.45   **2.** 2.6
3. 2.09   **4.** 11.19
5. 3.35
6. Possible answer: You can draw grids to help add the decimal numbers. I split $1.49 into 1 and 0.49. I split $1.19 into 1 and 0.19. I split $2.32 into 2 and 0.32. I had four whole grids to show 1 + 1 + 2 = 4. I combined the rest onto one grid to show 0.49 + 0.19 + 0.32 = 1. I had five whole grids filled with nothing left over. So she spent $5.

Chapter 2
Lesson 6, page 28
1. 6.72   **2.** 6.24
3. 11.00   **4.** 9.50
5. 10.32   **6.** 11.02
7. 4.93   **8.** 8.35
9. 24.22   **10.** 25.63
11. 5   **12.** 17.89
13. 8.71   **14.** 263.24
15. 15.45
16. 26.05 pounds

Chapter 2
Lesson 7, page 29
1. 4.2; Answers will vary.; Answers will vary.
2. 3.64; Answers will vary.; Answers will vary.
3. 10.45; Answers will vary.; Answers will vary.
4. 10.77; Answers will vary.; Answers will vary.
5. 9.5; Answers will vary.; Answers will vary.

Chapter 2
Lesson 8, page 30
1. 1.05   **2.** 0.80
3. 1.50   **4.** 8.49
5. 2.25
6. 11.27 kilograms

Chapter 2
Lesson 9, page 31
1. 1.11   **2.** 3.30
3. 6.09   **4.** 18.77
5. 1.79   **6.** 8.78
7. 1.77   **8.** 7.26
9. 4.41   **10.** 7.92
11. 3.44   **12.** 12.07
13. 1.66   **14.** 4.50
15. 91.09   **16.** 11.20
17. 113.84 pounds

Chapter 2
Lesson 10, page 32
1. 1.53; Answers will vary.; Answers will vary.
2. 0.98; Answers will vary.; Answers will vary.
3. 3.1; Answers will vary.; Answers will vary.
4. 0.02; Answers will vary.; Answers will vary.
5. 59.64; Answers will vary.; Answers will vary.

Chapter 2
Lesson 11, page 33
1. Yes. Possible answer: $7.75 + $7.75 + $4.50 + $4.50 = $24.50 on roller coaster rides; $0.85 + $0.85 + $0.85 + $0.85 = $3.40 for corn. $24.50 + $3.40 = $27.90 which is the total amount of money they would spend. They have $30, so they have enough.
2. The bag with the tomato cans weighs less.; 3.51 pounds less
3. 13.37 miles

Chapter 2
Test, pages 34–35
1. 675   **2.** 276
3. 76; 76   **4.** 56.89
5. 529   **6.** 703
7. 1,166   **8.** 938
9. 93.47   **10.** 890
11. 12.47   **12.** 150.5
13. 88.3; Answers will vary.; Answers will vary.
14. 311   **15.** 14.61
16. 286.95   **17.** 155
18. 41.31   **19.** 565.96
20. 10.1   **21.** 64.75
22. 35.4; Answers will vary.; Answers will vary.
23. the Brinker family
24. 1,210 beats per minute

Chapter 3

Chapter 3
Chapter 3
Lesson 1, page 36
1. $2 \times 4 = 4 \times 2 = 8$; Commutative
2. $4 \times 2 = 2 \times 4 = 8$; Commutative
3. 6; Identity
4. $3 \times 6 = 18$; Commutative
5. 8; Identity

Chapter 3
Lesson 2, page 37
1. 17; 20; 100; 1,700; Associative
2. 10; 37; 37; 407; Distributive
3. 51; 100; 1,400; Distributive
4. $(2 \times 4) \times 3 = 2 \times (4 \times 3) = 24$
5. Associative

Chapter 3
Lesson 3, page 38
1. 18;

$6 \times 3 = 18$; $18 \div 3 = 6$; $18 \div 6 = 3$
2. 20;

$4 \times 5 = 20$; $20 \div 4 = 5$; $20 \div 5 = 4$
3. 12;

$6 \times 2 = 12$; $12 \div 2 = 6$; $12 \div 6 = 2$
4. 2;

$1 \times 2 = 2$; $2 \div 1 = 2$; $2 \div 2 = 1$

Chapter 3
Lesson 4, page 39
1. 48; Possible answer: I changed 3 to 6 and 16 to 8. Then I multiplied $6 \times 8 = 48$.

2. 105; Possible answer,
   I multiplied 5 × 20 = 100.
   Then I added 5 × 1 = 5.
   100 + 5 = 105
3. 220; Possible answer: I
   doubled 55 = 110. Then I
   doubled 110 = 220.
   4 × 55 = 220
4. 675; Possible answer: I
   multiplied 200 × 3 = 600,
   20 × 3 = 60, 5 × 3 = 15;
   600 + 60 + 15 = 675
5. 93   6. 210
7. 552   8. 168
9. 847   10. 500
11. 1,104   12. 1,818

Chapter 3
Lesson 5, page 40
1. 2,500; Possible answer:
   I took 1 from 51 and
   added it to 49, making
   the problem 50 × 50 =
   2,500.
2. 2,236; Possible answer:
   I multiplied 43 × 2 ones
   = 86. I multiplied 43 × 5
   tens = 2,150 .I added 86
   and 2,150 = 2,236
3. 713   4. 1,054
5. 5,600   6. 722
7. 660   8. 5,950
9. 3,287   10. 1,845
11. 144 peaches

Chapter 3
Lesson 6, page 41
1. 7,599   2. 1,386
3. 2,499   4. 9,799
5. 15,717   6. 10,665
7. 19,614   8. 7,605
9. 5,625   10. 7,383
11. 10,234   12. 62,300
13. 26,362
14. 1,872 yards

Chapter 3
Lesson 7, page 42
1. 10 × 10 × 10; 49,000
2. 10 × 10; 100; 4,900
3. 10; 52.5
4. 10 × 10 × 10; 0.525
5. 1.024
6. 177,600
7. 22.5
8. exponent; moves to the
   right
9. power; by 10; less;
   moves to the left

Chapter 3
Lesson 8, page 43
For 1–6, strategies may
include: I used an array, a
model, or an equation.

1. 21   2. 52
3. 23   4. 17
5. 17   6. 83

Chapter 3
Lesson 9, page 44
1. 13   2. 12
3. 24   4. 3
5. 23   6. 14
7. 3   8. 6
9. 19   10. 6
11. 14   12. 7
13. Possible answer: To find
    58 ÷ 8, I drew an array.

Chapter 3
Lesson 10, page 45
1. 14 R1   2. 12 R4
3. 16 R2   4. 5 R2
5. 4 R4   6. 9 R3
7. 2 R4   8. 5 R3
9. 9 R4   10. 5 R2
11. 20 R3   12. 8 R1

Chapter 3
Lesson 11, page 46
1. 11 R14   2. 12 R6
3. 11 R12   4. 31 R15
5. 12 R50   6. 46 R24
7. 81 R31   8. 12 R64
9. 46 R24
10. Possible answer: The
    remainder should never
    be equal to or greater
    than the divisor. In
    this case, the quotient
    should be increased
    by 1 so that there is a
    remainder of only 2:
    17 – 15 = 2.

Chapter 3
Lesson 12, page 47
1. 28 elephant stamps;
   Solution: 4 leopard + 16
   tiger + 28 elephant = 48
   stamps.
2. The quotient will have a
   remainder of 1.

Chapter 3
Test, pages 48–49
1. 10; 37; 444; Distributive
2. 51; 100; 1,900;
   Distributive
3. 40; 8 × 5 = 40; 40 ÷ 5 =
   8; 40 ÷ 8 = 5
4. 63; 9 × 7 = 63; 63 ÷ 7 =
   9; 63 ÷ 9 = 7
5. 456   6. 7,200
7. 484   8. 5,226
9. 1,060   10. 52,500
11. 33,000   12. 10.1
13. 7.47   14. 0.506
15. 9 8-foot lengths, 6 feet
    left over

16. 7 rafts, 4 vests extra
17. 152.2 pounds
18. $3,375.00
19. 16 moose; Solution:
    4 bear + 16 moose +
    30 eagle = 50 photos
20. 20
21. No; He needs $43.50 to
    buy everything on his list
    and the blanket.

Chapter 4

Chapter 4
Lesson 1, page 50
Your estimates may vary.
1. 0.80; 0.84; Possible
   answer: I used a
   hundredths grid.
2. more than 0.30, less
   than 0.35; 0.33; Possible
   answer: Multiply 1 tenth
   × 3; add product to 1
   hundredth × 3.
3. 1; 1.00; Possible answer:
   I used place value.
4. a little more than 3; 3.06;
   Possible answer: I used
   hundredths grids.
5. little less than 4; 3.84;
   Possible answer: I used
   hundredths grids.
6. less than 5; 4.00;
   Possible answer: I used
   place value.

Chapter 4
Lesson 2, page 51
Your estimates may vary.
1. > 4.0; 4.2
2. < 12; 11.79
3. > 16; 17.76
4. > 32; 32.96
5. > 28; 28.07
6. > 12; 12.36
7. > 15, < 18; 16.62
8. < 7; 6.86
9. > 49; 49.21
10. > 6; 6.24
11. $8.34

Chapter 4
Lesson 3, page 52
Your estimates may vary.
1. < 1,800; 1,776
2. > 10,000; 10,240
3. 11,000; 11,000
4. 990,000; 990,000
5. exponent; product; right

Chapter 4
Lesson 4, page 53
1. 0.04; 0.04
2. < 0.30; 0.28

3. 0.24; 0.24
4. 0.30; 0.30
5. 0.72; 0.72
6. < 0.04; 0.004

Chapter 4
Lesson 5, page 54
1. 0.3; 2.4 ÷ 0.3 = 8; 0.3 × 8
   = 2.4; 8 × 0.3 = 2.4
2. 30; 6 ÷ 30 = 0.2; 30 × 0.2
   = 6; 0.2 × 30 = 6
3. .22; 2.2 ÷ .22 = 10; .22 ×
   10 = 2.2; 10 × 2.2 = 22
4. 20; 2 ÷ 20 = 0.1; 0.1 × 20
   = 2; 20 × 0.1 = 2
5. 20; 8 ÷ 20 = 0.4; 0.4 × 20
   = 8; 20 × 0.4 = 8
6. 0.32; 1.6 ÷ 0.32 = 5; 0.32
   × 5 = 1.6; 5 × 0.32 = 1.6
7. greater than

Chapter 4
Lesson 6, page 55
Your estimates may vary.
1. 0.2; 0.2
2. > 2; 2.20
3. 0.7; 0.7
4. > 60; 70
5. > 1; 1.3
6. > 1; 1.6
7. > 35; 50

Chapter 4
Lesson 7, page 56
1. 10; 10; 0.001; 0.0015
2. 10; 10; 0.009
Your estimates may vary.
3. < 1; 0.1776
4. > 0.01; 0.01024

Chapter 4
Lesson 8, page 57
Your estimates may vary.
1. 200; 200
2. > 2,000; 2,200
3. > 8; 8.1
4. > 800; 810
5. > 1,300; 1,330
6. 0.16; 0.16
7. interest ÷ 9,000 = 0.01
   or $9,000 × 0.01 =
   interest; $90

Chapter 4
Lesson 9, page 58
1. 12 baskets; 2 × 3 = 6
   baskets; 6 baskets × 2 =
   12 baskets
2. 9 goats; year 5: 144;
   year 4; 144 ÷ 2 = 72;
   year 3: 72 ÷ 2 = 36; year
   2: 36 ÷ 2 = 18; year 1: 18
   ÷ 2 = 9
3. 12 years old; 44 ÷ 2 =
   22; 22 – 10 = 12

Answers

Chapter 4
Test, pages 59–60
Your estimates may vary.
1. > 2; 2.6
2. > 12; 13.5
3. > 100; 105.6
4. > 12; 12.72
5. > 10; 10.05
6. < 0.10; 0.09
7. 0.24; 0.24
8. 0.24; 0.24
9. 0.48; 0.48
10. 0.48; 0.48
11. 0.2; 0.2
12. > 2; 2.2
13. 0.7, 0.7
14. > 5; 5.1
15. > 1.0; 1.3
16. 30; 30
17. > 2,000; 2,200
18. > 60; 70
19. > 700; 710
20. > 1,000; 1,200
21. > 10,000; 10,660
22. > 1,000; 1,195
23. 149.2; 149.2
24. 81.52; 81.52
25. < 0.1; 0.01999
26. > 0.11; 0.1155
27. less than 4; 3.5; Possible answer: I used place value.
28. 50; $5 \div 50 = 0.1$; $0.1 \times 50 = 5$; $50 \times 0.1 = 5$
29. 13.7×10^2 or 1, 370; $[(34.4 - 6.2 - 2.7 - 4.6 - 7.2) \times 10^2] = (34.4 - 20.7) \times 10^2 = 13.7 \times 10^2$
30. 2,000 bottles; 84 cases; $2,000 \div 24 = 83$ R8
31. 10,000; 21,000 pounds; $0.3 \times 7 = 2.1$; $2.1 \times 10^4 = 2.1 \times 10,000 = 21,000$; 4.8 ounces; $16 \times 0.3 = 4.8$

Chapter 5

Chapter 5
Lesson 1, page 61
1. $(10)^3 - (200) \times (3) + 10^2 =$; $1,000 - 200 \times (3) + 100 =$; $1,000 - 600 + 100 =$; $400 + 100 = 500$; $(15 - 5)^3 - (40 \times 5) \times (8 - 5) + 10^2 = 500$
2. $(100) \times (5) + (5)^2 - 10^1 =$; $100 \times (5) + 25 - 10 =$; $500 + 25 - 10 =$; $525 - 10 = 515$; $(25 \times 4) \times (9 - 4) + (10 - 5)^2 - 10^1 = 515$

Chapter 5
Lesson 2, page 62
1. $27 - 3 = 24$
2. $8 + 40 \times 6 = 8 + 240 + 248$
3. $(5 \times 2) \div 4 = 10 \div 4 = 2$ R2
4. $(35) - (4)^2 = 35 - 16 = 19$
5. $3^2 \times (10) + 7 - (6) = 9 \times 10 + 7 - 6 = 90 + 7 - 6 = 97 - 6 = 91$
6. $4 + 4 \times (5 \times 2) \div 4 = 4 + 4 \times (10) \div 4 = 4 + 40 \div 4 = 4 + 10 = 14$
7. $(8) + (10)^2 = 8 + 100 = 108$
8. Possible answer: $(6 \times 4) + (12 - 8) + 2^2 = 32$

Chapter 5
Lesson 3, page 63
1. $\{[3(20) + 27] \times 4\}$; $\{[60 + 27] \times 4\} = \{87 \times 4\}$; $\{87 \times 4\} = \{348\} = 348$; 348
2. $3\{[2(9) + 9] - 24\} = 3\{[18 + 9] - 24\}$; $3\{[27] - 24\}$; $3\{3\} = 9$; 9
3. $[(20 + 8) \div 14]^2 = [(28) \div 14]^2$; $[2]^2 = 4$; 4

Chapter 5
Lesson 4, page 64
1. $12 - (3 \times 2^2)$
2. $22 - (18 \div 2)$
3. $3^2 \times (5 - 2)$
4. $(6 \times 9) + 3$
5. Square the sum of 4 and 2, and then multiply by the product of 5 and 2.
6. Multiply the product of 5 and 3 by 6, and then subtract 50.
7. Multiply 3 squared by 2 squared, and then add the product of 5 and 4.
8. <; $(45 \div 9) \times 7 = 5 \times 7 = 35$; $24 \times 3 - 30 = 72 - 30 = 42$; $35 < 42$

Chapter 5
Lesson 5, page 65
1. 0.047; The difference between numbers is 0.020; $0.027 + 0.020 = 0.047$
2. Table should contain the following info:
   Bid 1: $2,000
   Bid 2: $2,500
   Bid 3: $3,000
   Bid 4: $3,500
   Bid 5: $4,000
   Bid 6: $4,500
   Bid 7: $5,000
   Bid 8: $5,500

3. 1 penny, 3 nickels, 4 dimes: $(4 \times 10) + (3 \times 5) + (1 \times 1) = 40 + 15 + 1 = 56$; or 2 quarters and 6 pennies = 56 $(2 \times 25) + (6 \times 1) = 56$

Chapter 5
Test, pages 66–67
1. $(10)^3 - (100) \times (4) + 10^2 =$; $1,000 - 100 \times (4) + 100 =$; $1,000 - 400 + 100 =$; $600 + 100 = 700$; $(15 - 5)^3 - (20 \times 5) \times (8 - 4) + 10^2 = 700$
2. $(150) \times (4) + (5)^2 - 10^1 =$; $150 \times (4) + 25 - 10 =$; $600 + 25 - 10 =$; $625 - 10 = 615$; $(30 \times 5) \times (6 - 2) + (10 - 5)^2 - 10^1 = 615$
3. $(7 \times 3) \div 9 = 21 \div 9 = 2$ R3
4. $61 - 14 = 47$
5. $12 + 20 \times 8 = 12 + 160 = 172$
6. $5 + 2^2 \times (10) \div 5 = 5 + 4 \times (10) \div 5 = 5 + 40 \div 5 = 4 + 8 = 12$
7. $(6) + (4)^2 = 6 + 16 = 22$
8. $2^2 \times 3^2 + 6 - (32) = 4 \times 9 + 6 - 32 = 36 + 6 - 32 = 42 - 32 = 10$
9. $\{[3(10) + 25] \times 2\}$; $\{[30 + 25] \times 2\} = \{55 \times 2\}$; $\{55 \times 2\} = \{110\} = 110$; 110
10. $3\{[2(9) + 8] - 20\} = 3\{[18 + 8] - 20\}$; $3\{[26] - 20\} = 3\{6\}$; $3\{6\} = 18$; 18
11. $[(30 + 4) \div 17]^2 = [(34) \div 17]^2$; $[2]^2 = 4$; 4
12. $15 - (4 \times 2^2)$
13. $54 - (36 \div 6)$
14. $7^2 \times (8 - 6)$
15. $(12 \times 15) + 3$
16. Square the sum of 4 and 2, and then subtract 16.
17. Multiply the product of 6 and 7 by 8, and then subtract 50.
18. Multiply 3 squared by 2 squared, and subtract the product from 48.
19. Multiply the product of 9×2 by 7, and then subtract 2 squared.
20. $(125) - 0 = 125$
21. $\{[4 \times (25) + 27] \times 4\} = \{[100 + 27] \times 4\} = \{127 \times 4\} = 508$
22. 1.259; The difference between numbers is 0.4; $0.859 + 0.4 = 1.259$
23. $(6 + 2)^2 - 30$
24. $3.87; 3.99 - (3.99 \times 0.03) = 3.99 - 0.12 = 3.87$;
25. 5; $21 - 6 = 15$ hours for Saturday and Sunday combined; 10 hours Sunday, 5 hours Saturday

Chapter 6

Chapter 6
Lesson 1, page 68
1. 4, 8, 12, 16, 20, 24, 28; 5, 10, 15, 20, 25, 30; 20
2. 8, 16, 24, 32, 40; 6, 12, 18, 24, 30; 24; 24
3. 4, 8, 12, 16, 20, 24, 30; 6, 12, 18, 24, 30; 12
4. 5, 10, 15, 20, 25, 30; 6, 12, 18, 24, 30; 30

Chapter 6
Lesson 2, page 69
1. $\frac{8}{16}$ 2. $\frac{4}{12}$
3. $\frac{4}{20}$ 4. $\frac{10}{15}$
5. $\frac{9}{12}$ 6. $\frac{12}{16}$
7. $\frac{12}{20}$ 8. $\frac{10}{24}$
9. $\frac{24}{30}$ 10. $\frac{35}{49}$

Chapter 6
Lesson 3, 70
1. $\frac{2}{4} = \frac{1}{2}$ 2. $\frac{7}{9}$
3. $\frac{4}{7}$ 4. $\frac{7}{31}$
5. $\frac{3}{9} = \frac{1}{3}$ 6. $\frac{13}{39} = \frac{1}{3}$
7. $\frac{33}{23} = 1\frac{10}{23}$
8. $\frac{10}{8} = 1\frac{2}{8} = 1\frac{1}{4}$
9. $\frac{10}{15} = \frac{2}{3}$ 10. $\frac{6}{9} = \frac{2}{3}$
11. $\frac{3}{18} = \frac{1}{6}$

Chapter 6
Lesson 4, page 71
1. $\frac{2}{4} = \frac{1}{2}$ 2. $\frac{8}{16} = \frac{1}{2}$
3. $\frac{3}{7}$ 4. $\frac{2}{8} = \frac{1}{4}$
5. $\frac{6}{16} = \frac{3}{8}$ 6. $\frac{17}{37}$
7. $\frac{10}{31}$ 8. $\frac{1}{9}$

9. $\frac{21}{53}$ 10. $\frac{5}{29}$

11. $\frac{2}{8} = \frac{1}{4}$ 12. $\frac{7}{15}$

13. $\frac{11}{35}$ 14. $\frac{5}{15} = \frac{1}{3}$

15. $\frac{13}{77}$ 16. $\frac{2}{24} = \frac{1}{12}$

17. $\frac{5}{65} = \frac{1}{13}$ 18. $\frac{4}{16} = \frac{1}{4}$

Chapter 6
Lesson 5, page 72

1. $\frac{5}{20} + \frac{4}{20} = \frac{9}{20}$

2. $\frac{4}{18} + \frac{3}{18} = \frac{7}{18}$

3. $\frac{9}{24} + \frac{8}{24} = \frac{17}{24}$

4. $\frac{13}{14}$

5. $\frac{19}{15} = 1\frac{4}{15}$

6. $\frac{15}{56}$ 7. $\frac{13}{42}$

Chapter 6
Lesson 6, page 73

1. $1\frac{3}{12} + \frac{2}{12} = 1\frac{5}{12}$

2. $1\frac{4}{6} + \frac{3}{6} = 1 + \frac{7}{6} = 1 + 1\frac{1}{6} = 2\frac{1}{6}$

3. $4\frac{9}{24} + 2\frac{4}{24} = 4 + 2 + \frac{13}{24} = 6\frac{13}{24}$

4. $2\frac{2}{18} + \frac{15}{18} = 2\frac{17}{18}$

5. $6\frac{13}{24}$ 6. $2\frac{7}{10}$

7. $\frac{13}{14}$ 8. $3\frac{14}{15}$

9. $12\frac{5}{56}$ 10. $7\frac{31}{42}$

Chapter 6
Lesson 7, page 74

1. $\frac{4}{18} - \frac{3}{18} = \frac{1}{18}$

2. $\frac{9}{24} - \frac{8}{24} = \frac{1}{24}$

3. $\frac{3}{6} - \frac{2}{6} = \frac{1}{6}$

4. $\frac{15}{20} - \frac{4}{20} = \frac{11}{20}$

5. $\frac{1}{14}$ 6. $\frac{5}{12}$

7. $\frac{2}{99}$ 8. $\frac{1}{42}$

Chapter 6
Lesson 8, page 75

1. $1\frac{3}{12} - \frac{2}{12} = 1\frac{1}{12}$

2. $1\left(\frac{4}{6} - \frac{3}{6}\right) = 1 + \frac{1}{6} = 1\frac{1}{6}$

3. $1\frac{21}{24} - 1\frac{12}{24} = \frac{9}{24}$

4. $3\frac{9}{24} - 2\frac{6}{24} = 1\frac{3}{24} = 1\frac{1}{8}$

5. $2\frac{2}{18} - 1\frac{15}{18} = 1\frac{2}{18} = 1\frac{1}{9}$

6. $3\frac{5}{10} - 2\frac{2}{10} = 1\frac{3}{10}$

7. $3\frac{7}{14} - 1\frac{6}{14} = 2\frac{1}{14}$

8. $2\frac{5}{15} - 1\frac{6}{15} = \frac{14}{15}$

9. $5\frac{36}{42} - 4\frac{5}{32} = 1\frac{1}{42}$

10. $6\frac{21}{24} - 1\frac{16}{24} = 5\frac{5}{24}$

Chapter 6
Lesson 9, page 76

1. $30 + 10 = 40$

2. $40 - 30 = 10$

3. $70 - 20 = 50$

4. $30 + 20 = 50$

5. $1 + 1 = 2$

6. $1 - \frac{1}{2} = \frac{1}{2}$

7. $1 - 0 = 1$

8. $0 + \frac{1}{2} = \frac{1}{2}$

Chapter 6
Lesson 10, page 77

1. $\frac{1}{12} + \frac{1}{16}$; $\frac{7}{48}$ are either apple or berry: $\frac{1}{12} + \frac{1}{16} = \frac{4}{48} + \frac{3}{48} = \frac{7}{48}$

2. $\frac{1}{3} + \frac{1}{12}$; $\frac{5}{12}$ of the boxes are either orange or apple juice: $\frac{4}{12} + \frac{1}{12} = \frac{5}{12}$

3. 1 box is $\frac{1}{48}$ of the whole number of juice boxes. Change each fraction to one with a denominator of 48 and the numerator will be the number of boxes Orange, $\frac{1}{3} = \frac{16}{48}$; 16 boxes; Apple; $\frac{1}{12} = \frac{4}{48}$; 4 boxes; Berry; $\frac{1}{16} = \frac{3}{48}$; 3 boxes.

Chapter 6
Lesson 11, page 78

1. $3\frac{5}{6}$: $3\frac{1}{3} + \frac{1}{2} = 3\frac{2}{6} + \frac{3}{6} = 3\frac{5}{6}$; Diagrams will vary.

2. 65 boys, 80 girls; Diagrams will vary.

Chapter 6
Test, pages 79–80

1. 30. 2. 56
3. 12 4. 42
5. 120 6. 54
7. 10 8. 15
9. $\frac{11}{22}$ 10. $\frac{8}{20}$
11. $\frac{15}{35}$ 12. $\frac{8}{12}$
13. $\frac{9}{15}$ 14. $\frac{14}{21}$
15. $\frac{2}{4} = \frac{1}{2}$ 16. $\frac{7}{9}$
17. $\frac{1}{7}$ 18. $\frac{5}{5} = 1$
19. $\frac{4}{17}$ 20. $\frac{2}{8} = \frac{1}{4}$

21. $\frac{8}{56} - \frac{7}{56} = \frac{1}{56}$

22. $1\frac{7}{14} + 1\frac{6}{14} = 3\frac{5}{14}$

23. $3\frac{10}{15} + 2\frac{9}{15} = 5\frac{19}{15} = 6\frac{4}{15}$

24. $5\frac{7}{42} - 4\frac{6}{42} = 1\frac{1}{42}$

25. $1\frac{1}{2} + 1\frac{2}{3} = 1\frac{3}{6} + 1\frac{4}{6} = 2\frac{7}{6} = 3\frac{1}{6}$

26. $3\frac{8}{20} - 2\frac{5}{20} = 1\frac{3}{20}$

27. $8\frac{8}{56} - 7\frac{7}{56} = 1\frac{1}{56}$

28. $5\frac{7}{42} + 4\frac{6}{42} = 9\frac{13}{42}$

29. 744 square feet; Sample solution: Step 1: Find the area of the pool + deck = 42 × 32 = 1,344 square feet. Step 2: Find the area of the pool alone: 30 × 20 = 600 square feet. Step 3: Find the area of deck alone: 1,344 − 600 = 744 square feet. Diagram should show a smaller rectangle labeled 20 x 30 feet inside of a larger rectangle labeled 26 x 36 feet

30. $\frac{3}{4}$ block

31. $\frac{5}{6} - \frac{1}{2} = \frac{1}{3}$

32. 14.75 cm: 1.75 cm every week after the first week.

33. 12: 25 − 7 = 18 hours for Friday and Sunday combined; 12 hours Sunday, 6 hours Saturday

34. Romero: swimming pool; Hill: forest preserve; Johnson: zoo

Chapter 7

Chapter 7
Lesson 1, page 81

1. 5 2. 8
3. 6
4. Possible answer:

5. Possible answer:

Chapter 7
Lesson 2, page 82

1. Each will receive $1\frac{1}{4}$ pounds of cheese: $\frac{5}{4} = 1\frac{1}{4}$.

2. Possible answer: No; $\frac{21}{4} = 5\frac{1}{4}$; There cannot be $\frac{1}{4}$ of a sheep in a field.

3. $\frac{1}{2}$ pound of blueberries in each container

4. $\frac{82}{5} = 16$ with 2 left over; Possible answer: Two people will have to carry 17 bags each.

Chapter 7
Lesson 3, page 83

1. One half of the drama class are boys wearing glasses: $\frac{3}{4} \times \frac{2}{3} = \frac{6}{12} = \frac{1}{2}$

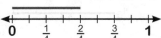

Answers

2.

$\frac{1}{6}$ of the actors' costumes need to be replaced: $\frac{2}{3} \times \frac{1}{4} = \frac{2}{12} = \frac{1}{6}$

3.

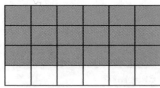

$\frac{5}{8}$ of the students in drama class take the same bus to school: $\frac{5}{6} \times \frac{3}{4} = \frac{15}{24} = \frac{5}{8}$

Chapter 7
Lesson 4, pages 84–85

1. $\frac{4}{8} = 2$ 2. $\frac{6}{2} = 3$

3. $\frac{12}{2} = 6$ 4. $\frac{9}{3} = 3$

5. $\frac{5}{10} = \frac{1}{2}$ 6. $\frac{3}{3} = 1$

7. $\frac{24}{3} = 8$ 8. $\frac{32}{4} = 8$

9. $\frac{45}{9} = 5$ 10. 8

11. 9 12. 6

13. 5 14. 4

15. $8\frac{2}{3}$ 16. 9

17. 18 18. 12

19. 6 20. 2 cans

21. $8\frac{1}{3}$ hours

Chapter 7
Lesson 5, pages 86–87

1. $\frac{1}{32}$ 2. $\frac{1}{12}$

3. $\frac{1}{4}$ 4. $\frac{2}{6} = \frac{1}{3}$

5. $\frac{3}{12} = \frac{1}{4}$ 6. $\frac{4}{20} = \frac{1}{5}$

7. $\frac{2}{9}$ 8. $\frac{3}{25}$

9. $\frac{12}{20} = \frac{3}{5}$ 10. $\frac{10}{36} = \frac{5}{18}$

11. $\frac{6}{75}$ 12. $\frac{9}{64}$

13. $\frac{16}{25}$ 14. $\frac{10}{18} = \frac{5}{9}$

15. $\frac{10}{48} = \frac{5}{24}$ 16. $\frac{14}{84} = \frac{1}{6}$

17. $\frac{3}{15} = \frac{1}{5}$ 18. $\frac{3}{44}$

19. $\frac{3}{40}$ 20. $\frac{3}{32}$

21. $\frac{3}{20}$

22. $\frac{8}{15}$ are green; $\frac{4}{5} \times \frac{2}{3} = \frac{8}{15}$

23. $\frac{1}{6}$ is copper; $\frac{2}{3} \times \frac{3}{8} = \frac{6}{24} = \frac{1}{6}$

24. $\frac{5}{12}$ is recycled; $\frac{1}{2} \times \frac{5}{6} = \frac{5}{12}$

25. $\frac{3}{32}$ was used in the kitchen; $\frac{3}{8} \times \frac{1}{4} = \frac{3}{32}$

26. $\frac{2}{3} \times \frac{4}{5} = \frac{8}{15}$

Chapter 7
Lesson 6, page 88

1. $\frac{12}{4} = 3$ square units;

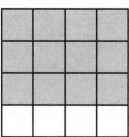

2. $\frac{6}{3} = 2$ square units;

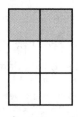

3. $\frac{4}{8} = 2$ square units;

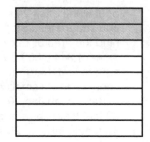

4. $\frac{1}{3}$ square unit;

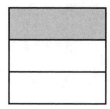

5. $\frac{10}{3} = 3\frac{1}{3}$ sq yd

6. $\frac{16}{5} = 5\frac{1}{5}$ sq m

Chapter 7
Lesson 7, page 89

1. twice; 6
2. twice; great; twice; 5
3. four times; great; four times greater; 6
4. half as great
5. five times as great
6. five times smaller
7. twice as large
8. three times smaller

Chapter 7
Lesson 8, page 90–91

1. more than 4; $\frac{24}{5} = 4\frac{4}{5}$

2. more than 5; $\frac{35}{6} = 5\frac{5}{6}$

3. equal to 7; $7 \times 1 = 7$

4. more than 7; $\frac{35}{3} = 11\frac{2}{3}$

5. less than 4; $20/6 = 3\ 1/3$

6. less than 9; $63/8 = 7\ 7/8$

7. less than 11; $\frac{165}{16} = 10\frac{5}{16}$

8. less than 9; $\frac{27}{16} = 1\frac{11}{16}$

9. less than 9; $\frac{63}{16} = 3\frac{15}{16}$

10. less than 9; $\frac{135}{16} = 8\frac{7}{16}$

11.

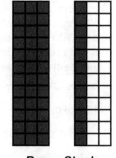

Repro Steel:
$12 \times \frac{4}{3} = \frac{48}{3} = 16$

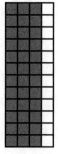

Secondhand Steel:
$12 \times \frac{3}{4} = \frac{36}{4} = 9$

Chapter 7
Lesson 9, page 92

1. $3\frac{5}{6}$ bushels; $4\frac{3}{5} \times \frac{5}{6} = \frac{23}{5} \times \frac{5}{6} = \frac{115}{30} = \frac{23}{6} = 3\frac{5}{6}$

Chapter 7
Lesson 10, page 93

1. 3–5 videos

Videos	Tally	Number of Friends
0–2	///	3
3–5	####	5
6–8	////	4
9–11	///	3

2. 0–2 and 9–11; 3

Chapter 7
Test, pages 94–95

1. $4 \div 2 = 2$
2. $5 \div 3 = 1\ R2$
3. $8 \div 5 = 1\ R3$
4. $12 \div 3 = 4$

5. $\frac{6}{8} = \frac{3}{4}$ 6. $\frac{4}{2} = 2$

7. $\frac{12}{3} = 4$ 8. $\frac{18}{3} = 6$

9. $\frac{15}{10} = 1\frac{1}{2}$ 10. $\frac{12}{4} = 3$

11. $\frac{1}{32}$ 12. $\frac{1}{72}$

13. $\frac{1}{4}$ 14. $\frac{4}{9}$

15. $\frac{6}{12} = \frac{1}{2}$ 16. $\frac{12}{20} = \frac{3}{5}$

17. more than 7; $\frac{49}{6} = 8\frac{1}{6}$

18. less than 7; $\frac{42}{7} = 6$

19. more than 9; $\frac{36}{3} = 12$

20. less than 9; $\frac{117}{14} = 8\frac{5}{14}$

21. less than 11; $\frac{11}{99} = \frac{1}{9}$

22. less than 11; $\frac{1,078}{99}$ = $10\frac{8}{9}$

23. 15 students

24. About 22 students

25. More drink 3–4 glasses of milk a day

Milk	Tally	Number of Friends
1–2	ℍℍ II	7
3–4	ℍℍ ℍℍ III	13

26. $3 \times \frac{2}{3} = 2$

27. $\frac{9}{10}$ acre

28. 16 feet tall: $3\frac{1}{5} \times 3 = \frac{16}{5} \times 3 = \frac{48}{3} = 16$

Chapter 8

Chapter 8
Lesson 1, page 96

1. $\frac{1}{10}$ 2. $\frac{1}{12}$

Chapter 8
Lesson 2, page 97–98

1. $\frac{1}{10}$; $\frac{1}{10} \times 5 = \frac{5}{10} = \frac{1}{2}$

2. $\frac{1}{8}$; $\frac{1}{8} \times 4 = \frac{4}{8} = \frac{1}{2}$

3. $\frac{1}{9}$; $\frac{1}{9} \times 3 = \frac{3}{9} = \frac{1}{3}$

4. $\frac{1}{24}$; $\frac{1}{24} \times 6 = \frac{6}{24} = \frac{1}{4}$

5. $\frac{1}{15}$; $\frac{1}{15} \times 3 = \frac{3}{15} = \frac{1}{5}$

6. $\frac{1}{15}$; $\frac{1}{15} \times 5 = \frac{5}{15} = \frac{1}{3}$

7. $\frac{1}{16}$; $\frac{1}{16} \times 4 = \frac{4}{16} = \frac{1}{4}$

8. $\frac{1}{35}$; $\frac{1}{35} \times 7 = \frac{7}{35} = \frac{1}{5}$

9. $\frac{1}{18}$; $\frac{1}{18} \times 3 = \frac{3}{18} = \frac{1}{6}$

10. $\frac{1}{4}$; $\frac{1}{4} \times 2 = \frac{2}{4} = \frac{1}{2}$

11. $\frac{1}{9}$; $\frac{1}{9} \times 3 = \frac{3}{9} = \frac{1}{3}$

Chapter 8
Lesson 3, page 99

1. 35   2. 15
3. 12

Chapter 8
Lesson 4, page 100–101
Drawings will vary.

1. 9; $9 \times \frac{1}{3} = \frac{9}{3} = 3$

2. 20; $20 \times \frac{1}{4} = \frac{20}{4} = 5$

3. 9; $9 \times \frac{1}{3} = \frac{9}{3} = 3$

4. 14; $14 \times \frac{1}{2} = \frac{14}{2} = 7$

5. 15; $15 \times \frac{1}{3} = \frac{15}{3} = 5$

6. 36; $36 \times \frac{1}{9} = \frac{36}{9} = 4$

7. 20; $20 \times \frac{1}{4} = \frac{20}{4} = 5$

8. 21; $21 \times \frac{1}{3} = \frac{21}{3} = 7$

9. 24; $24 \times \frac{1}{4} = \frac{24}{4} = 6$

Chapter 8
Lesson 5, page 102

1. 3 miles   2. 30 days
3. $35.91

4. $\frac{1}{12}$ of the remaining water

Chapter 8
Test, pages 103–104

1. $\frac{1}{8}$ 2. $\frac{1}{12}$

3. 16   4. 30

5. $\frac{1}{15} \times 5 = \frac{1}{3}$

Your drawings will vary.

6. $\frac{1}{20}$; $\frac{1}{20} \times 5 = \frac{5}{20} = \frac{1}{4}$

7. $\frac{1}{18}$; $\frac{1}{18} \times 6 = \frac{6}{18} = \frac{1}{3}$

8. $\frac{1}{14}$; $\frac{1}{4} \times 7 = \frac{7}{14} = \frac{1}{2}$

9. 24; $24 \times \frac{1}{4} = \frac{24}{4} = 6$

10. 20; $20 \times \frac{1}{5} = \frac{20}{5} = 4$

11. 10; $10 \times \frac{1}{2} = \frac{10}{2} = 5$

12. 12 weeks

Chapter 9

Chapter 9
Lesson 1, page 105

1. 11.6 km

2. 50 cm

3. No; Possible answer: I can convert centimeters to millimeters. The box is 280 millimeters tall but only 320 millimeters wide. The model is wider than that.

4. 2,574 cm   5. 1,400 m

6. 22.8 m

Chapter 9
Lesson 2, page 106

1. 1,890 g   2. 11.8 kg

3. It will take two trips; Possible answer: I can convert kilograms to metric tons. The pipes are 19.45 metric tons. They can ship 10 tons the first trip and 9.45 tons the second trip.

4. 113,000 mg

5. 0.06 g   6. 11,900 kg

Chapter 9
Lesson 3, page 107

1. 70,000 mL

2. 0.57 L

3. The artist does not have enough paint; Possible answer: I can convert deciliters to liters and then add. The artist has a total of 25.1 liters of paint. That's not enough.

4. 2.2 dL

5. 20 glasses

6. 256,000 mL

Chapter 9
Lesson 4, page 108

1. 90 minutes

2. 489 seconds

3. 54.75 hours

4. 3,640 weeks

5. 2 years

6. 37,230 days

Chapter 9
Lesson 5, page 109

1. 12°C

2. 16°C; 30°C

3. 33°C

4. 34°C; 9°C

Chapter 9
Lesson 6, page 110

1. 43°F

2. 22°F; 61°F

3. 103°F

4. 96°F; 77°F

Chapter 9
Lesson 7, page 111

1. the oak tree

2. 136 yards taller

3. Yes; Possible answer: I can convert feet to inches. The closet is 84 inches long and 48 inches wide. The bicycle is smaller than that so it should fit.

4. 27,456 ft   5. 8 in.

6. 730.33 yd or $730\frac{1}{3}$ yd or 730 yards, 1 foot

Chapter 9
Lesson 8, page 112

1. 384 oz

2. 2,400 lbs heavier

3. Yes; Possible answer: I converted ounces to pounds. Mona has 13.75 pounds of apples and 10 pounds of bananas. Altogether, they weigh 23.75 pounds, which the basket can hold without breaking.

4. 16,000 lb   5. 10,000 lb

6. 76.8 oz

Chapter 9
Lesson 9, page 113

1. 8 qt   2. 1 c

3. 128 qt; Possible answer: I found the sum of the gallons and then converted the answer to quarts. If all three gas tanks are filled, they would have 12 + 10 + 10 = 32 gallons in all. If I multiply 32 by 4, that's 128 quarts.

4. 10 pt

5. 32 pt; 16 qt

6. 15 gal

Chapter 9
Lesson 10, page 114

1. 152 km; You do not need to know that Katie walks about half as much to solve the problem.

2. 3.82 liters; You do not need to know that 19 people came to the picnic to solve the problem.

3. 158.13 kg; You do not need to know that $\frac{2}{3}$ of the food they eat is fish to solve the problem.

4. The problem cannot be solved. You need to know how many students are in the class.

Chapter 9
Lesson 11, page 115

1. 24 c or $1\frac{1}{2}$ gal

2. 20.15 lb

3. Possible answers: 567, 319; 591,736; 579,163: other answers are possible, but all must have 5 as the first digit.

4. 854.3 mi

Answers

5. 160 c

6. $7\frac{5}{12}$ in.

1. 18°C
2. 18°C; 39°C
3. 31°F
4. 52°F; 65°F
5. Yes; Possible answer: I can convert millimeters to centimeters. The box is 34 centimeters wide and 20 centimeters tall. The bowl is smaller than that. It should fit.
6. 27,280 m
7. 1,340 g
8. 64,000 mg
9. 125 mL
10. 498 seconds
11. 162 inches taller
12. 499,840 yd
13. 48 qt
14. 6 pt
15. The problem cannot be solved. You need to know how many gallons the store sold on Tuesday.
16. $75\frac{1}{2}$ blocks; You do not need to know that he stopped at a park for 30 minutes to solve the problem.

Chapter 10

Chapter 10
Lesson 1, page 118
1. Possible answer: Point A, Point G

$\overline{BD}$ $\overline{GF}$

2. Possible answer: BD, GF

$\overleftrightarrow{BE}$ $\overleftrightarrow{AF}$

3. Possible answer: BE and AF

$\overleftrightarrow{AF}$ $\overleftrightarrow{FE}$

4. Possible answer: answer: AF and FE

$\overleftrightarrow{GD}$ $\overleftrightarrow{BE}$

5. Possible answer: GD and BE

Chapter 10
Lesson 2, page 119
1. acute; 84°
2. obtuse; 147°
3. obtuse; 169°
4. right; 90°
5. straight; 180°
6. acute; 21°

Chapter 10
Lesson 3, page 120
1. equilateral triangle, acute triangle
2. right triangle, acute triangle equilateral
3. scalene triangle, obtuse triangle
4. scalene triangle, obtuse triangle
5. isosceles triangle, acute triangle
6. equilateral triangle, acute triangle

Chapter 10
Lesson 4, page 121
1. trapezoid
2. square or parallelogram or rhombus or rectangle
3. rhombus or parallelogram
4. rectangle or parallelogram
5. Possible answer: I do not agree. Squares have four right angles because rectangles have four right angles, and all squares are rectangles. Not all rhombi have right angles.

Chapter 10
Lesson 5, page 122
1. A, D, and E are polygons; B and C are not polygons.
2. triangle
3. hexagon
4. quadrilateral
5. Possible answer: A circle is not a polygon because a circle is not made up of line segments.

Chapter 10
Lesson 6, page 123
1. irregular
2. regular
3. regular;

Flowchart:
column 1 below scalene: isosceles, equilateral
column 2 ending with acute: Obtuse; right
column 3, 1 pair, trapezoid; bottom row: square, Yes; rhombus, No; trapezoid, or parallelogram

Chapter 10
Lesson 7, page 124
1. 36 m²
2. 5,505.64 km²
3. 7,060.76 mm²
4. 4,096 cm²
5. $3,862\frac{11}{16}$ in.²
6. $\frac{16}{25}$ mi²
7. The square has the greater area. The rectangle's area measures 255 m². The square's area measures 256 m².

Chapter 10
Lesson 8, page 125
1. 8   2. 27
3. 24   4. 18
5. 15   6. 16

Chapter 10
Lesson 9, page 126
1. 45 cm³   2. 64 ft³
3. 8 cm × 2 cm × 3 cm = 48 cm³

Chapter 10
Lesson 10, page 127
1. (4 × 2) × 2 = 16 m³
2. (11 × 11) × 11 = 1,331 in.³
3. (8 × 4) × 2 = 64 cm³
4. (16 × 8) × 3 = 384 yd³

Chapter 10
Lesson 11, page 128
1. 7,182 mm³
2. 2,262 ft³
3. 17,576 mi³
4. 4,913 in.³
5. 588 yd³
6. 1,755 mi³

Chapter 10
Lesson 12, page 129
1. 147 mm³
2. 4,200 ft³
3. 1,764 km³
4. 24,480 m³

Chapter 10
Lesson 13, page 130
1. 2,400,000 cm³
2. 8 ft   3. 18 boxes
4. 4 m   5. 120 in.³

Chapter 10
Test, pages 131–132
1. Possible answer: $\overline{AB}$, $\overline{DC}$
2. $\overleftrightarrow{AE}$ and $\overleftrightarrow{BD}$
3. Possible answer: $\overrightarrow{AE}$ and $\overleftrightarrow{CE}$
4. acute; 22°
5. obtuse; 164°
6. straight; 180°
7. first triangle
8. circle, third shape
9. first shape
10. 65 m²
11. 289 cm²
12. 27 mm³
13. 216 mi³
14. 576 km³
15. 7,533 yd³
16. 2,112 ft³
17. 572 cm³

Chapter 11

Chapter 11
Lesson 1, page 133
1. Line plot should be numbered from 2 to 5 with quarter intervals labeled, and should show the following info:

1 X above $2\frac{1}{4}$,

2 X's above $3\frac{1}{4}$

3 X's above $3\frac{1}{2}$

1 X above $3\frac{3}{4}$

1 X above 4

1 X above $4\frac{1}{4}$

2. $2\frac{1}{4}$ kg

3. A box should be drawn around $3\frac{1}{4} - 3\frac{3}{4}$.
2 ovals should be drawn. One around $2\frac{1}{4} - 3$, and another around $4\frac{1}{2} - 5$.

4. 2

5. $3\frac{1}{2}$ kg

Chapter 11
Lesson 2, page 134
1. $2\frac{1}{4}$ in.   2. $86\frac{3}{4}$ in.
3. 4; 170 in.

4. $125\frac{1}{2}$ in.

5. $41\frac{1}{4}$ in.

6. $42\frac{9}{16}$ in.

Chapter 11
Lesson 3, page 135

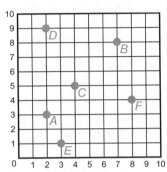

Chapter 11
Lesson 4, page 136
1. square;

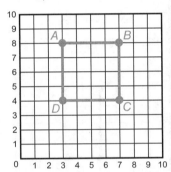

2. trapezoid;

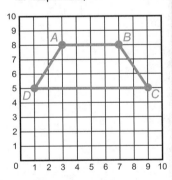

3. triangle or obtuse scalene triangle;

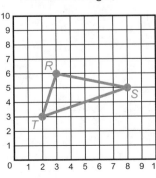

Chapter 11
Lesson 5, page 137
1. triangle; (2, 9), (9, 2), (2, 2);

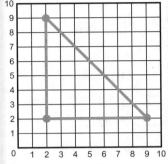

2. Possible answer: (1, 1), (1, 3), (4, 1) (4,3);

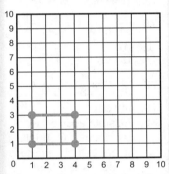

3. parallelogram; *A* (4, 6), *B* (2, 3), *C* (9, 6), *D* (7, 3);

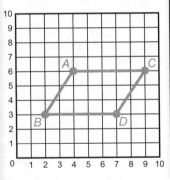

Chapter 11
Lesson 6, page 138
1. left-hand table: 4; 5; 6; 7; 8; right-hand table: 7; 8; 9; 10; 11; Possible answer: Each number in the second pattern is 3 more than the number in the first pattern.
2. left-hand table: 8; 7; 6; 5; 4; right-hand table: 5; 4; 3; 2; 1; Each number in the second pattern is 3 less than the number in the first pattern.

3. left-hand table: 0, 4, 8, 12, 16; right-hand table: 0, 2, 4 ,6, 8; Except for zero, each number in the second pattern is half the number in the first pattern.

Chapter 11
Lesson 7, page 139
1. top table: 2; 3; 4; 5; 6; bottom table: 3; 4; 5; 6; 7;

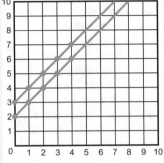

2. top table: 7; 6; 5; 4; 3; bottom table: 8; 7; 6; 5; 4;

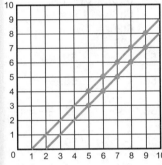

3. top table: 2; 4; 6; 8; 10; bottom table: 3; 4; 5; 6; 7;

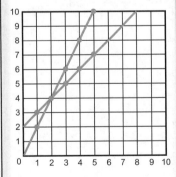

Chapter 11
Lesson 8, page 140
1.

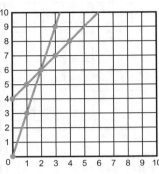

Chapter 11
Test, pages 141–142

1. $3\frac{2}{10}$ m or $3\frac{1}{5}$ m

2. 2

3. $3\frac{8}{10}$ m or $3\frac{4}{5}$ m

4. $10\frac{6}{10}$ m or $10\frac{3}{5}$ m

5. $\frac{7}{10}$ m

6. $3\frac{7}{10}$ m

7–11.

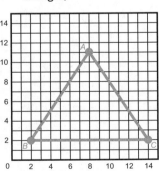

12. triangle or equilateral triangle;

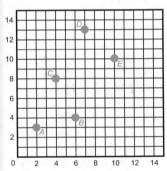

Answers

13. Rectangle;

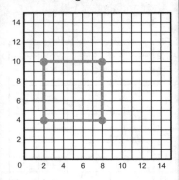

Posttest
Pages 143–148

1. 0.039   **2.** 17.871

3. 84,652,000,034

4. seven billion, one hundred eleven million, six hundred fifty-eight thousand, three hundred fourteen

5. twelve billion, five hundred million, five hundred thousand, five hundred

6. >   **7.** <

8. =   **9.** 7.12

10. 5   **11.** 6.1

12. 6.53   **13.** 900

14. 744   **15.** 4.88

16. 79,071   **17.** 9

18. 68.87   **19.** 387

20. 0.8   **21.** 159

22. 177.05   **23.** 74,749

24. 854   **25.** 2.9

26. 5.01 ounces

27. 5; 12; 120; 180; Distributive Property

28. 8; 30; 60; 480; Associative Property

29. 768   **30.** 126

31. 3,249   **32.** 9,168

33. 3,480 pounds

34. 168,500   **35.** 590,000

36. 9,870   **37.** 5 R3

38. 45   **39.** 9 R5

Your estimates may vary.

40. > 12; 12.08

41. < 30; 29.7

42. > 18; 18.42

43. < 1; 0.1588

44. > 0.08; 0.08723

45. > 29; 29.84

46. > 130; 137

47. > 77,000; 77,430

48. < 16,000; 15,850

49. 12.75 quarts;
$0.75 + (1.5 \times 8) =$
$0.75 + 12 = 12.75$

50. $(10)^3 - (60) \times (5) + 10^4$;
$1,000 - 60 \times 5 + 10,000$;
$1,000 - 300 + 10,000$;
$700 + 10,000 \; 10,700$;
$10,700$

51. $6\{[4(8) + 3] - 15\} =$
$3\{[32 + 3] - 15\}$;
$6\{[35] - 15\}$;
$6\{20\} = 120$;
120

52. 109   **53.** 78

54. $\dfrac{3}{6} = \dfrac{1}{2}$ **55.** $\dfrac{4}{9}$

56. $\dfrac{6}{8} = \dfrac{3}{4}$ **57.** $\dfrac{23}{18} = 1\dfrac{5}{18}$

58. $\dfrac{3}{8}$ **59.** $\dfrac{31}{63}$

60. $7\dfrac{7}{40}$ **61.** $12\dfrac{1}{24}$

62. $7\dfrac{1}{3}$ **63.** $1\dfrac{15}{28}$

64. $\dfrac{6}{9}$ **65.** $\dfrac{10}{20}$

66. $\dfrac{7}{28}$ **67.** $\dfrac{14}{16}$

68. 4

69. $\dfrac{105}{8} = 13\dfrac{1}{8}$

70. $\dfrac{7}{7} = 1$

71. 30 pink roses

72. $\dfrac{9}{14}$ are leafy greens

73. 336 days

74. $\dfrac{27}{8} = 3\dfrac{3}{8}$ sq ft

75. $\dfrac{5}{3} = 1\dfrac{2}{3}$ sq m

Your drawings may vary.

76. $\dfrac{1}{54}$; $\dfrac{1}{54} \times 9 = \dfrac{9}{54} = \dfrac{1}{6}$

77. $\dfrac{1}{14}$; $\dfrac{1}{14} \times 7 = \dfrac{7}{14} = \dfrac{1}{2}$

78. 72; $72 \times \dfrac{1}{8} = \dfrac{72}{8} = 9$

79. 440 cm

80. Possible answer: No. I can convert Gabrielle's height to inches. She is 51 inches tall. That is 3 inches shorter than the height she needs to be.

81. 1,097 days

82. 87°F; higher that average; 12° higher

83. Possible answer: Point F, Point A, and Point C

84. Possible answer: $\overleftrightarrow{AB}$ and $\overleftrightarrow{EF}$

85. Possible answer: $\overline{AE}$, $\overline{CF}$, and $\overline{EB}$

86. acute; 81°

87. obtuse; 135°

88. blue oval; green incomplete hexagon

89. Possible answer: I do not agree. A parallelogram has 2 pairs of parallel sides. A trapezoid has 1 pair of parallel sides.

90. 864 cm³

91. A line plot should be drawn from 1 to 3 with $\dfrac{1}{4}$ intervals labeled, and should show the following info:
1 X over $1\dfrac{1}{2}$, with a circle around it
1 X over 3
1 X over $3\dfrac{1}{4}$
3 X's over $3\dfrac{1}{4}$

92. rectangle;

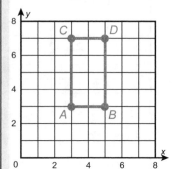